Turtles OF ALABAMA

Philip Henry Gosse as a young man of twenty-nine, the year of his return to England from Alabama, painted by his brother, William Gosse. (1839, watercolor on ivory, courtesy of the National Portrait Gallery—London)

INAUGURAL SPONSORS

Birmingham Audubon Society
Mary Carolyn Gibbs Boothby
Lida Inge Hill
Fay Belt Ireland
Henry S. Lynn Jr.
Medical Management Plus, Inc.
Dr. Connie S. and Dr. James A. Pittman Jr.
Elberta Gibbs Reid
Frances B. and James W. Shepherd
Becky H. and William E. Smith Jr.
Virginia B. and William M. Spencer III
Ann A. and Donald B. Sweeney Jr.
Dr. Cameron M. and Judge J. Scott Vowell
Alice M. and N. Thomas Williams
Harriett Harton Wright
Louise A. and John N. Wrinkle

Philip Henry Gosse (1810–1888) was an English naturalist and illustrator who spent eight months of 1838 on the Alabama frontier, teaching planters' children in Dallas County and studying the native flora and fauna. Years after returning to England, he published the now-classic *Letters from Alabama: Chiefly Relating to Natural History,* with twenty-nine important black-and-white illustrations included. He also produced, during his Alabama sojourn, forty-nine remarkable watercolor plates of various plant and animal species, mainly insects, now available in *Philip Henry Gosse: Science and Art in "Letters from Alabama" and "Entomologia Alabamensis."*

The Gosse Nature Guides are a series of natural history guidebooks prepared by experts on the plants and animals of Alabama and designed for the outdoor enthusiast and ecology layman. Because Alabama is one of the nation's most biodiverse states, its residents and visitors require accurate, accessible field guides to interpret the wealth of life that thrives within the state's borders. The Gosse Nature Guides are named to honor Philip Henry Gosse's early appreciation of Alabama's natural wealth and to highlight the valuable legacy of his recorded observations. Look for other volumes in the Gosse Nature Guides series at http://uapress.ua.edu.

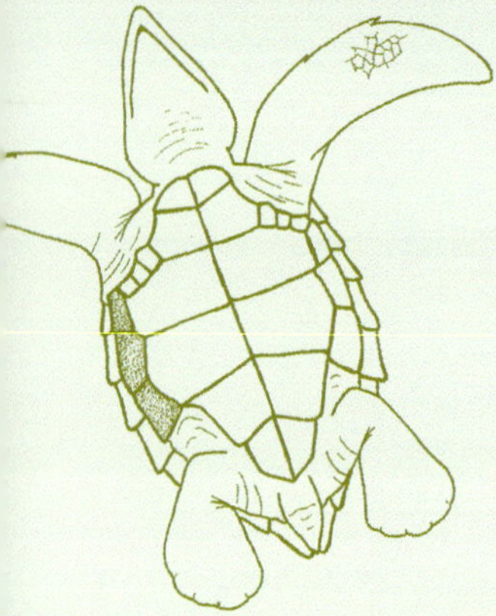

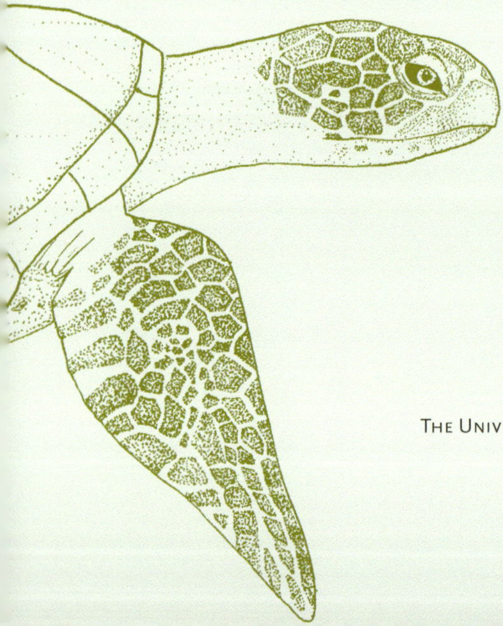

The University of Alabama Press Tuscaloosa

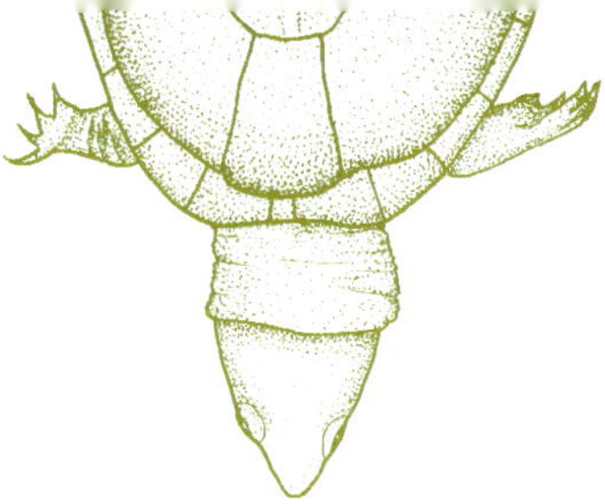

Turtles OF ALABAMA

CRAIG GUYER

MARK A. BAILEY

ROBERT H. MOUNT

WITH LINE DRAWINGS BY CLAIRE L. FLOYD

The University of Alabama Press
Tuscaloosa, Alabama 35487-0380
uapress.ua.edu

Typeface: Scala Pro and Scala Sans Pro

Manufactured in China
Cover photograph: Adult female Northern Map Turtle, *Graptemys geographica*; courtesy of Alan Cressler
Cover and interior design: Michele Myatt Quinn
∞
The paper on which this book is printed meets the minimum requirements of American National Standard for Information Sciences—Permanence of Paper for Printed Library Materials, ANSI Z39.48-1984.

Cataloging-in-Publication data is available from the Library of Congress.
ISBN: 978-0-8173-5806-8
E-ISBN: 978-0-8173-8817-1

Publication made possible in part by generous donations from Gary R. Mullen, in memorial to his friend George W. Folkerts, and Carole Sommers, in memorial to her friend Mary Ward Brown.

Yet, quiet and apparently desolate as this somber lake is, it is the congenial home of some animals, and those not to be despised by man. It is true, if you approach noisily, kicking the stumps and breaking the twigs, you see nothing; nor will you be any the wiser if you move about the brink talking and laughing. But sit down on a log a few paces within the shadow of the forest, and remain quite silent, keeping your eyes on the surface of the pool, and especially on the fallen trunks that project from it. In a few minutes a little black head peeps from the mantle of green incumbent weed, and a Tortoise creeps noiselessly out, and takes up his position on one of the rotten logs. Glance over the pool; every log is tenanted by one or more of the same silent reptiles, not one of which was there a moment ago. But lo! While you look, another and another and another—nay, scores are crawling up upon the logs, so that in a quarter of an hour you may count them by the hundreds, and others are still rising. They are of various sizes; some are as large as the crown of your hat, others are tiny creatures, not bigger that a half-crown piece, and of all intermediate dimensions.

—Philip Henry Gosse, *Letters from Alabama,* Letter IV (Dallas County, June 10, 1838)

Contents

Abbreviations

ADCNR	Alabama Department of Conservation and Natural Resources
AUM	Auburn University Museum of Natural History
CITES	Convention on International Trade in Endangered Species
ESA	Endangered Species Act of 1973, as amended
IUCN	International Union for Conservation of Nature
NOAA	National Oceanic and Atmospheric Administration
SWAP	Alabama's State Wildlife Action Plan (2005)
USFWS	United States Fish and Wildlife Service

Turtles OF ALABAMA

Introduction

This book is designed to update the turtle fauna described in Mount's (1975) comprehensive volume on the reptiles and amphibians of Alabama. Our treatment represents the first in a series of volumes that will cover each major taxonomic group described in Mount's seminal work. Alabama covers one of the most species-rich regions in the Northern Hemisphere and, for groups like turtles, rivals the diversity that might be observed anywhere on Earth. Here, we provide a modern description of that diversity.

Much has happened in organismal biology since 1975. Significant field studies, especially of Alabama's threatened and endangered species, have been performed. The field of systematics has reemerged as a primary goal of biological sciences, and this has been coupled with a healthy debate on species concepts (e.g., Frost and Hillis 1990). This debate has expanded the focus of studies of speciation from tests of reproductive isolation (e.g., Carr and Stancyk 1975) to discovery of diagnostic features indicative of unique lineages on phylogenetic trees (e.g., Butler et al. 2011). These changes have increased the known diversity of the state and pointed out new directions for research that are likely to continue to expand Alabama's known turtle fauna.

Our main goal is to describe each species found within Alabama by providing diagnostic features, summarizing key life history variables, and indicating conservation efforts and management tools designed to maintain each species. To reach this goal, we first list the taxa currently known from the state and then present climatic, geologic, and geographic features that shape turtle diversity. We end our introductory material by outlining the information selected to characterize each species account. Our target audience remains the same as that for Mount (1975). We aim to enlighten people who are interested in the natural history of their local biota because we know these people will develop responsible attitudes toward the role that humans play in sustaining the Earth's ecosystems. Moreover, those with knowledge of natural history and a willingness to experience nature have a vast new world full of opportunities for soul-enriching experiences that we have had as biologists and that we hope to generate for others. This publication is a compromise of sorts, prepared for use by the layman as well as for the serious student of southeastern herpetology. The life history accounts are focused on providing information of interest to the former, while detailed information regarding taxonomy is presented for the latter.

The Turtle Fauna of Alabama

Indigenous Species

The classification scheme that follows is used to organize the native turtles of Alabama. These are taxa that are thought to have evolved within the state or to have dispersed there without the assistance of humans. Changes to systematic biology since the publication of Mount (1975) have generated a growing number of taxonomic problems. Generally, these are associated with a desire for taxonomic groupings that are monophyletic (groups in which members are all more closely related to each other than any member is to a species outside of the group) and a desire for restricting a proliferation of named groups associated with monophyletic taxonomies. In order to reach these goals we have adopted some of the philosophy argued by de Queiroz and Gautier (1992), who advocate reducing a reliance on taxonomic levels of the Linnean hierarchy in favor of generating indented lists of increasingly more restricted monophyletic groups. Even in such taxonomies, species are identified as binomials, with a genus name identifying a group of closely related species, and the specific epithet identifying a particular one of those species. The species name includes both the genus and the specific epithet, simultaneously generating a unique name for each species and identifying it as part of a more inclusive taxonomic group. In addition to this convention, we retain the level of family as a useful taxonomic category because this level is so heavily entrenched in the taxonomic literature and because the content of reptile families has remained relatively consistent. We have avoided use of terms associated with levels of the Linnean hierarchy above the level of the family because these vary substantially among schemes, and the choice of a term for these levels (e.g., superfamily versus suborder) is a matter of personal choice rather than providing any increased understanding of biology.

Our classification scheme uses the format of an indented list, starting with the group Testudines, the radiation that contains all living and fossil turtles. At the first level of indentation are the seven living families known from Alabama. These are listed in the same order as they appear in the keys because attempts to discover the evolutionary relationships of these families have not converged on a consistent phylogenetic tree. Within families we also list species and subspecies as they are presented in the keys, as follows:

- TESTUDINES
  - Dermochelyidae
    - *Dermochelys coriacea*—Leatherback Sea Turtle
  - Cheloniidae
    - *Caretta caretta*—Loggerhead Sea Turtle
    - *Lepidochelys kempii*—Kemp's Ridley
    - *Eretmochelys imbricata imbricata*—Atlantic Hawksbill
    - *Chelonia mydas*—Green Sea Turtle
  - Trionychidae
    - *Apalone calvata*—Gulf Coast Smooth Softshell
    - *A. mutica*—Midland Smooth Softshell
    - *A. ferox*—Florida Softshell
    - *A. spinifera spinifera*—Eastern Spiny Softshell
    - *A. spinifera aspera*—Gulf Coast Spiny Softshell
  - Testudinidae
    - *Gopherus polyphemus*—Gopher Tortoise
  - Chelydridae
    - *Chelydra serpentina*—Eastern Snapping Turtle
    - *Macrochelys temminckii*—Alligator Snapping Turtle
  - Kinosternidae
    - *Kinosternon baurii*—Striped Mud Turtle
    - *K. subrubrum hippocrepis*—Mississippi Mud Turtle
    - *K. subrubrum subrubrum*—Eastern Mud Turtle
    - *Sternotherus odoratus*—Eastern Musk Turtle (Stinkpot)

S. carinatus—Razor-backed Musk Turtle
S. depressus—Flattened Musk Turtle
S. minor—Loggerhead Musk Turtle
S. peltifer—Stripe-necked Musk Turtle

Emydidae

Terrapene carolina—Eastern Box Turtle
T. bauri—Florida Box Turtle
Malaclemys terrapin pileata—Mississippi Diamond-backed Terrapin
Deirochelys reticularia reticularia—Eastern Chicken Turtle
Chrysemys dorsalis—Southern Painted Turtle
C. picta picta—Eastern Painted Turtle
Graptemys geographica—Northern Map Turtle
G. barbouri—Barbour's Map Turtle
G. pulchra—Alabama Map Turtle
G. ernsti—Escambia Map Turtle
G. ouachitensis ouachitensis—Ouachita Map Turtle
G. nigrinoda nigrinoda—Northern Black-knobbed Sawback
G. nigrinoda delticola—Southern Black-knobbed Sawback
Trachemys scripta elegans—Red-eared Slider
T. scripta scripta—Yellow-bellied Slider
Pseudemys alabamensis—Alabama Red-bellied Turtle
P. concinna—River Cooter
P. floridana—Florida Cooter

Introduced Species

Because of increased trade in vertebrates, the establishment and expansion of non-indigenous species has become an increasing problem in maintaining native North American faunas (Romagosa et al. 2009). In the case of turtles, three native species found in Alabama (*Chrysemys picta, Apalone spinifera, Trachemys scripta elegans*) have successfully colonized other areas outside the state, largely because these species are sold as pets, food, or scientific subjects. Based on a specimen discovered near the confluence of the Raft River with Grand Bay in Baldwin County during 1994, it appears that the Ouachita Map Turtle has expanded its range within Alabama. The completion of the Tenn-Tom Waterway in 1984, a canal created by the US Army Corps and Engineers that connects the Tennessee River to the Tombigbee

River, may have provided an avenue for the expansion of this turtle species. A similar expansion of the Midland Smooth Softshell Turtle might be expected via this same pathway.

Fortunately, no non-native turtle species has invaded Alabama. However, four non-native species have been found in Alabama, all terrestrial box turtles or tortoises and all failing to establish viable populations. Two were observed in the area of Birmingham, the Red-footed tortoise (*Chelonoidis carbonarius*) and Horsfield's Tortoise (*Testudo horsfieldii*); two others were observed in Auburn, the Snake-eating Turtle (*Cuora flavomarginata*) and the African Spurred Tortoise (*Centrochelys sulcata*). All four likely represent released or escaped pets, and these species are unlikely ever to form viable populations because of low rates of colonization. Because Alabama has such a rich turtle fauna, its invasion by non-native turtle species may be unlikely as virtually all niches available to turtles may already be filled by native forms.

Taxonomic Changes and Problems

Our taxonomic list includes forty lineages (species, subspecies or genetic clades), a number that exceeds any other state in the nation and that, at first glance, represents a modest change from the thirty-eight listed in Mount (1975). However, this number masks relatively significant changes to the state's turtle fauna. All taxa on our list have valid scientific names, including the two mitochondrial lineages of Alligator Snapping Turtles known from Alabama based on Thomas et al. (2014). However, because Thomas et al. (2014) failed to provide diagnostic morphological features distinguishing *Macrochelys apalachicolae* from *M. temminckii*, we retain the traditional taxonomy by considering these to be one species, *M. temminckii*. We add four additional taxa. Three species, the Florida Box Turtle, the Striped Mud Turtle, and the Razor-backed Musk Turtle, are now documented from the state. Taxonomic separation of the Escambia Map Turtle from the Alabama Map Turtle accounts for the fourth addition. Six subspecies from Mount (1975) have been elevated to species status (see Appendix 1), altering our list from that previous effort. Finally, we have eliminated three taxa listed by Mount (1975). We now consider painted turtles from the Tennessee River in Jackson County that possess plastral patterns expected of the Midland Painted Turtle (*Chrysemys picta marginata*) to be Southern Painted Turtles, based on molecular work in Starkey et al. (2003), but displaying past evidence of hybridization with Eastern Painted Turtles. Following Butler et al. (2011), we eliminate the Gulf Coast Box Turtle (*Terrapene carolina major*) as a valid living taxon and restrict this name to the fossil lineage of giant box turtles of the Gulf Coast region. Based on that work we also consider box turtles of the Gulf Coast that possess three hind toes to belong to the Florida Box Turtle rather than the Three-toed Box Turtle (*Terrapene triunguis*).

Mount (1975) listed two taxa that are near Alabama's border and that might eventually be documented from the state. We now conclude that these are not present. The Yellow-blotched Sawback (*Graptemys flavimaculata*) is known to occur in some streams of the Pascagoula River system in Mississippi, but has not yet been found in streams of that system in Alabama, despite several attempts to discover it. The Escatawpa River is the largest of these, but it may be too small to provide sufficient food resources (mussels) to support *Graptemys*, a feature used to explain why no *Graptemys* are present in the Perdido River. Similarly, the Florida Red-bellied Turtle (*Pseudemys nelsoni*) occurs in the Apalachicola drainage of Florida, making it possible for this species to occur in the extreme southeastern corner of Alabama. However, the Chattahoochee River is home to dense populations of River Cooters that occupy that portion of the river forming the boundary between Alabama and Georgia. No suitable habitat appears to be present for the Florida Red-bellied Turtle, which occurs in broad, slow-moving, and even brackish waters of the mouths of rivers. Such habitat is only found along the coast of Florida.

Climate of Alabama

Because of its location, with a southern border along the Gulf Coast and a northern border along the southern extent of the Appalachian Mountains, the climate of Alabama is classified as Humid Subtropical (McKnight and Hess 2000). This climate is characterized by mild winters and hot, humid summers. Mean temperatures are warmer and more constant in the southern portion of the state than in the northern portion, both because of lower southern topography and a stronger influence of Gulf breezes in the south. Rainfall is distributed throughout the year because of cold, dry polar fronts moving against warm, moist, coastal air during autumn and winter, yielding intense thunderstorms, and moist, warm Gulf air moving north during spring and summer, rising over terrestrial areas, and generating afternoon rains. Rainfall is slightly increased along the coast of Alabama because of the increased moisture content of air associated with the Gulf of Mexico. Measurable snowfalls are exceedingly rare in the southern half of the state, and annual totals of more than 6 inches (150 mm) are seldom recorded even for the northernmost stations.

These typical patterns of weather are broken by annual occurrences of violent weather associated with tornadoes, mostly during spring, and hurricanes during late summer and fall. These storms can cause periods of intense rains that saturate soils and flood extensive areas. Such occurrences cause increased movement in some turtles, such as Gopher Tortoises. These storms also kill trees by tipping them up from the roots, snapping them off at the trunk, or severely stressing them from saltwater storm surges. Each of these add fallen logs and leaves to the floor of Alabama's forested habitats, thus providing sites used as refuges for species, such as Eastern Box Turtles, to avoid extreme temperatures or to serve as nest sites.

Alabama Geography

Alabama covers 52,419 square miles (135,765 km^2) of land and water in the southeastern United States and is divided into sixty-seven counties.

Essentially all of this area is habitable by turtles. Distributions of these organisms are affected most strongly by the river systems that provide habitat for most species. This association with drainage basins is so strong that geology and topography have little additional effect on explaining most turtle distributions within Alabama. Below, we summarize major geographic features of the state and discuss their effects on Alabama's turtle fauna.

River Basins

From a turtle's perspective, the most important feature of Alabama's geography is the diversity of rivers that drain its surface. Eight river systems are found in the state, all of which drain into the Gulf of Mexico.

Three of these are major rivers, being wide and long enough to present opportunities for diversification of aquatic organisms within them as well as challenges to terrestrial organisms attempting to cross them.

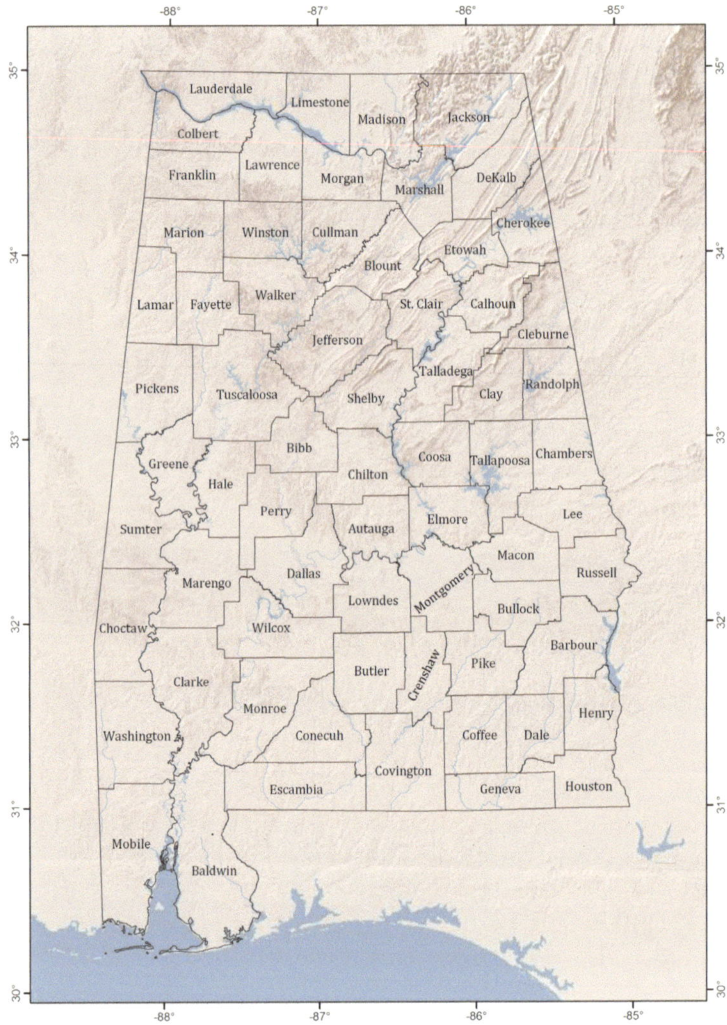

Alabama counties

The Tennessee River enters the northeast corner of the state, flows east to west, and exits the northwest corner. This river drains the northern one-eighth of the state and joins the Ohio River, exiting into the Gulf in Louisiana via the Mississippi River. However, the tortured pathway of the Tennessee River suggests that it changed its course from a Mesozoic path that took it southwestward down the Ridge and Valley formation to the Alabama River system (Appalachian River). In the Paleocene epoch (56 to 65 million years ago), erosion and tectonic events altered this ancient path by opening a channel from the Ridge

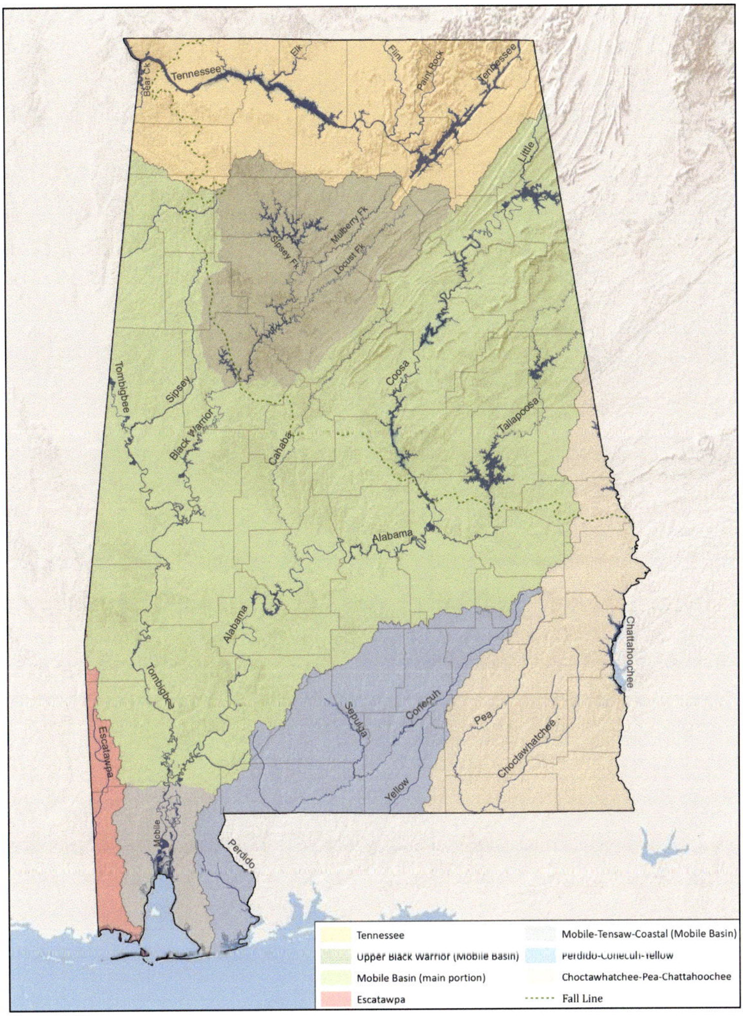

Alabama river drainages

and Valley formation into the Sequatchie Valley west of Chattanooga, following that valley southwestward to Guntersville, and then turning abruptly northwestward at Guntersville by cutting through the Pottsville sandstone formation into the Tuscumbia limestone. The Tennessee River takes a final turn at the Alabama-Mississippi-Tennessee border and eventually establishes a connection to the Ohio River (Mills and Kaye 2001). This unusual path and chain of historical events likely isolated some taxa in the Upper Tennessee River that share an ancient affinity with taxa in the Alabama River (e.g., Yellow-bellied

Above: Tennessee River from Painted Bluff, Marshall County, AL

Slider and Cumberland Slider [*Trachemys scripta troosti*]) and opened a pathway for taxa from the Mississippi River to enter the Tennessee River (e.g., Midland Smooth Softshell, Southern Painted Turtle, Ouachita Map Turtle).

Two additional major river basins are present in the state. One, the Chattahoochee, forms the southern half of Alabama's eastern border (draining six percent of the state), and the other, the Mobile, drains most of the state (sixty-four percent), entering the Gulf near Alabama's southwestern corner. These are ancient rivers that have remained in close association with upland regions of the southern Appalachians. However, their openings to the Gulf of Mexico have lengthened or shortened because of marine inundation or subsidence, altering their connectivity with smaller rivers during the 300 million years during which turtles have occupied this region. Independent evolution of sister taxa in these two drainages (e.g., Barbour's Map Turtle in the Chattahoochee and Alabama Map Turtle in the Mobile) suggest that these waters have been isolated from each other for exceptionally long time periods.

Five smaller rivers are also found in the state (see Map 2). One of these, the Escatawpa, originates in southwestern Alabama, draining the western portions of Mobile and Washington Counties (about three

Left: Chattahoochee River from I-85 bridge, Chambers County, AL

Below: Tensaw River from I-65 bridge, Baldwin County, AL

Escatawpa River at Lott Road Bridge, Mobile County, AL

Perdido River at Duck Place Road, Baldwin County, AL

percent of Alabama's surface area), but exits the state into Mississippi, where it opens to the Gulf in Pascagoula Bay. The Razor-backed Musk Turtle is found in Alabama only in this river.

The Perdido River drains Escambia and Baldwin Counties, opening into Perdido Bay. This short river drains only about one percent of Alabama and, from a herpetological perspective, is notable for lacking any member of the genus *Graptemys*.

Above: Yellow River, Covington County, AL

Left: Choctawhatchee River, Geneva County, AL

Two rivers, the Conecuh and Yellow, drain eight percent of Alabama in the south central counties of the state. These rivers leave the state, eventually entering Pensacola Bay in Florida. Receding shorelines likely connected these rivers over geological time, as evidenced by their sharing the Escambia Map Turtle.

Finally, the Choctawhatchee and Pea Rivers drain about six percent of southeastern Alabama and unite before entering Florida and, eventually, Choctawhatchee Bay. These short rivers appear to share a common history with the Chattahoochee River, because of the presence in all three of Barbour's Map Turtle, and with the Yellow and Conecuh Rivers, because of the common presence of the Escambia Map Turtle.

Important Geographic Units

From the perspective of terrestrial habitats, Alabama is divisible into ten meaningful physiographic units, which Mount (1975) termed herpetofaunal regions.

These units fall into two natural groupings: Coastal Plain units of the southern and western parts of the state that were created by ancient seashores, and Upland units that were formed by the southern end of the Appalachian Mountains. These two groupings have a distinct boundary, the Fall Line, where streams change from rocky, fast-flowing waterways of Upland units to sandy or muddy slow-moving waterways of the Coastal Plain.

The Coastal Plain in Alabama is distinctly belted and is physiographically more variable than it is in either Georgia or Mississippi. The belts are roughly crescent in shape and arc from Mississippi to Georgia. From south to north, they constitute the Lower Coastal Plain, Red Hills, Black Belt, and Fall Line Hills. Topographically these belts vary from flat to almost montane. The soils vary from acid sands and sandy loams, the dominant soils of the southern and northern belts, to clay sands, silts, and heavy, calcareous, alkaline types, soils that dominate regions of the central belts. The sandy soils are covered with pine forests, especially longleaf (*Pinus palustris*) in areas where native forests have been retained, and loblolly (*Pinus taeda*) or slash (*Pinus elliottii*) in areas where offsite pines have been planted. Those soils that are not primarily sandy allow for the formation of steep ravines that support dense hardwood forests and include upland areas with heavy soils that are covered by grasslands lacking a pine overstory.

Rocks and rock outcrops occur in some portions of the Coastal Plain, but seldom to the extent that they do in provinces above the Fall Line. Most streams of the Coastal Plain are fairly sluggish and have sand, silt, or gravel bottoms, but some flow over bedrock. Several of the streams have broad, low floodplains with sloughs and oxbows. Swampy habitats are fairly common, many of them having been created by beaver dams.

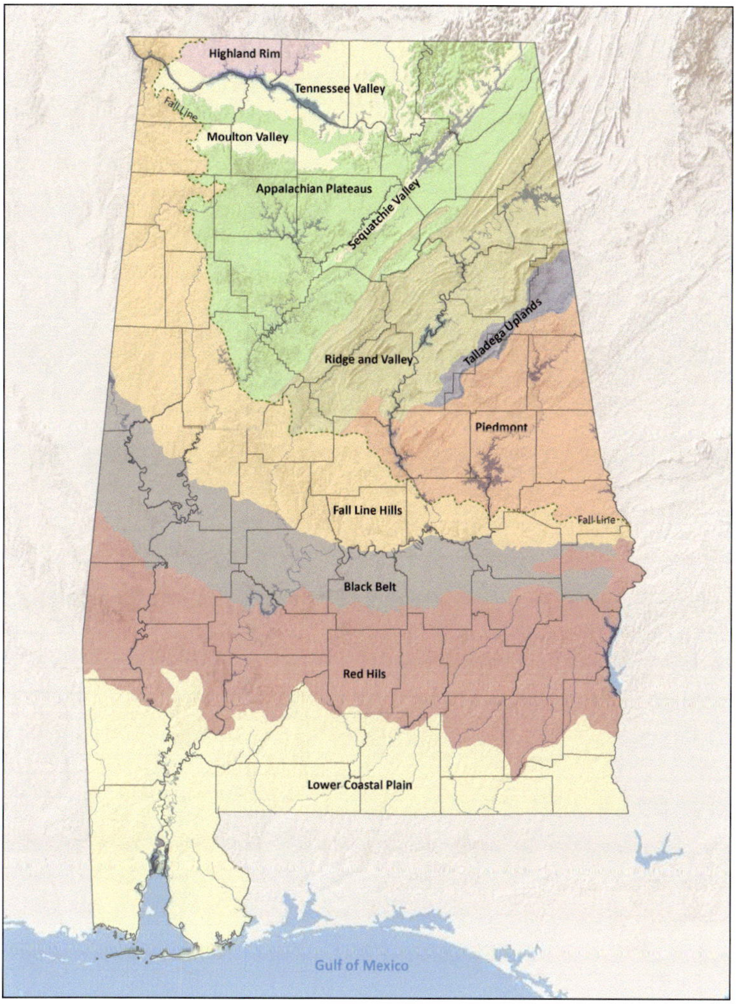

Major herpetofaunal regions of Alabama

Twenty-four lineages of turtles, constituting about 60 percent of the state's turtle lineages, occur exclusively or almost exclusively within the Coastal Plain (see Appendix 2). All five species of sea turtles, along with the Southern Black-knobbed Sawback, Mississippi Diamond-backed Terrapin, Alabama Red-bellied Turtle, Florida Box Turtle, Striped Mud Turtle, Mississippi Mud Turtle, Razor-backed Musk Turtle, and Florida Softshell, are restricted to the Lower Coastal Plain. This diversity peaks in and around Mobile Bay, where twenty-seven lineages may be found. There are no other examples of specialization of turtles in other regions of the Coastal Plain.

The Upland units comprise six regions located above the Fall Line. The southernmost of these regions is the Piedmont, a triangular formation located in east-central Alabama. This region is a portion of an ancient microcontinent (some parts of which are one billion years old) that collided with the eastern portion of North America, causing the uplift of the Appalachian Mountains. This latest uplift (350 million years ago) created a narrow mountainous region, the Talladega Uplands, which contains the tallest peaks in the state, along with the Ridge and Valley, a narrow southwest-to-northeast series of more or less uniformly folded terrain. Lying immediately above the Ridge and Valley region are the Appalachian Plateaus, a wide region that is a subdivision of the Cumberland Plateau, broad tabletop mountains of sandstone and limestone. The Upland regions are dissected by the Tennessee Valley, a broad expanse of relatively level, fertile land that lies along the Tennessee River, but that also includes bluffs and steep slopes on the east and south sides of the Tennessee Valley, where it meets the Appalachian Plateaus. The northernmost Upland region in Alabama is the Highland Rim, a region termed the Chert Belt by Mount (1975). Physiographically the Highland Rim is the southernmost subdivision of a vast province whose components are termed collectively the Interior Low Plateaus.

In general, the Upland regions have a limited effect on Alabama's turtle fauna. However, an exception to this rule is the Flattened Musk Turtle, a species restricted to the Upper Black Warrior drainage of the Appalachian Plateaus. Additionally, the Northern Map Turtle, a species of northern origin, reaches its southern limit in the Ridge and Valley of Alabama, and the Eastern Painted Turtle, a species of the Atlantic Coastal Plain, reaches its western limit in the Ridge and Valley and Piedmont of eastern Alabama. Finally, the Midland Smooth Softshell and Eastern Spiny Softshell, two species associated with the Mississippi River basin, are restricted to the Tennessee Valley of Alabama.

Species Accounts

The remainder of this book describes turtles as a major radiation of amniotes (terrestrial, egg-laying, or live-bearing vertebrates), each family of turtles found in Alabama, and each of the state's species within each family. Presentation of each family, genus, species, and subspecies is in the order of appearance within the keys provided rather than listed alphabetically. Important genetic variation supported by published analyses is discussed within each species account. Each species or subspecies account has distinct sections that we describe below.

Keys

Keys are tools designed to aid in identification of organisms. These tools present paired descriptions, one of which will conform to an individual organism of interest and the other of which will not. At the end of each consistent description is a number indicating the next couplet to be considered. This process of making dichotomous choices is followed until a final description identifying the organism of interest is reached. We include taxonomic keys for the turtles of Alabama and take the unusual step of dispersing these keys throughout the accounts rather than including a single key. We do this to place information close to sections of text for which the keys are most useful. Keys to the families of turtles appear at the end of the description of the group Testudines. When necessary, keys to genera are placed at the end of the description of each family, keys to species appear at the end of descriptions of each genus, and keys to subspecies are placed at the end of the Taxonomy section of the appropriate species account.

Names

The generic, specific, subspecific, and common names applied are, in most cases, those listed in Crother (2012). However, we have taken the liberty to elevate *Apalone mutica calvata* to species status because all current evidence supports this status and only taxonomic inertia has prevented the action previously.

Photographs

We have benefited from the talents of a large number of photographers. Where possible, we have selected images that show key features. When the location of the specimen photographed is known, we identify it.

Descriptions

Our descriptions are intended to provide sufficient information to enable the reader to distinguish a particular taxon from all others occurring within the state. Each description is based on a composite of specimens representing variation within Alabama and surrounding states. But, because nature is variable, it should be kept in mind that occasional individuals belonging to the described taxon will not conform to the descriptions presented here.

Alabama Distribution

In addition to a general statement describing the distribution of each taxon occurring within Alabama, a distribution map is included. For each map, we include diagonal hatching intended to delineate the presumed range at the state level. As these data are incomplete for most taxa, our estimated ranges should not be expected to be entirely accurate. Maps for species thought to occur statewide lack hatching. Known occurrences are presented as red dots on a Herpetofaunal Regions base map and readers may use these, along with the discussion in the text, as further clues to overall distribution. These red dots depict specific locations for (a) specimens the authors have examined, (b) photo-vouchered specimens submitted to the Alabama Herp Atlas Project (a web-based program of citizen science records stored at AUM), (c) occurrences documented in the databases of the Alabama Natural Heritage Program and/or ADCNR State Lands Division, Natural Heritage Section, and (d) literature records believed valid. Each record was georeferenced and plotted to the greatest possible precision using ArcGIS® software by Esri. Purple dots indicate specimens found displaying evidence of hybridization or intergradation with another species/subspecies. The state distribution maps also include insets showing the approximate distribution of each taxon within the continental United States or Gulf Coast. These are modified and customized from a number of previously published distribution maps, particularly those developed for Elliot et al. (2008).

Habits

Here we provide information on habitat selection, seasonal patterns of activity and reproduction, mating strategies of males and females, major diet items, and habitat associations. In general, this section is designed to summarize where and when each species is likely to be active and what activities make the species detectable by humans. Additionally, we describe the timing and duration of each major growth stage in the life cycle of each turtle species.

Conservation and Management

In this section we describe the current conservation status of each species in Alabama. Alabama's turtles are generally threatened by habitat loss and fragmentation, loss of natural community integrity, and direct persecution. Because conservation issues are likely to increase in the future, we summarize management activities that might imperil each species as well as those activities that are likely to enhance populations. Similar data are provided for species that have conservation status within the state, and for these we provide information on key public properties that will play crucial roles in long-term maintenance of Alabama's imperiled turtles. In developing its 2005 State Wildlife Action Plan, or SWAP, (Alabama Department of Conservation and Natural Resources, Division of Wildlife and Freshwater Fisheries 2005), the State of Alabama used the findings from its 2002 Nongame Symposium, which assembled scientific experts to compile the best data on Alabama's wildlife, and used those data to identify those species most in need of conservation action. The Third Nongame Symposium's Amphibian and Reptile Subcommittee identified eleven turtles as being of immediate conservation need (Priority 1 or 2, on a scale of 1 to 5 as described in Mirarchi et al. 2004), and we summarize the subcommittee's recommendation in each species account.

Taxonomy

We accept the concept that species are lineages that are discovered through careful analysis of variation in the characteristics of organisms. These discoveries arise from creation of phylogenetic trees built from character data. Under this species concept, any diagnosable terminal branch is sufficient to discover a new species. Additionally, we accept the concept that taxonomic groups at any level of classification should be monophyletic. In practice, ancestral species might survive through the branching process, generating some lineages that are difficult to diagnose (de Queiroz 1998). Such species present challenges for determining species boundaries, and the decisions that we make for the boundaries of Alabama's species undoubtedly will suffer from this challenge.

For turtles, color patterns, conformation of scutes, counts of scales, and shapes of appendages are external features that traditionally have been used to diagnose species, genera, and families. To these traditional characters we have added information from publications describing the mitochondrial and nuclear genomes. These characters have the advantage of allowing rapid development of data sets that are much larger than those based on morphology. The mitochondrial genome is inherited in offspring entirely from the female side of the family tree, while the nuclear genome captures information about gene flow associated with both parents. For this reason, phylogenetic trees based on the mitochondrial genome are not guaranteed to be concordant with those based on the nuclear genome. Nevertheless, published data on the mitochondrial genome often are assumed to carry phylogenetic information, in the absence of data to the contrary. Because these mitochondrial data are cheap to generate, are voluminous, and typically tell an evolutionary story that is consistent with the nuclear genome, we accept them as providing evidence of inter- and intraspecific evolution. These data are particularly important in phylogeographic studies, a field of biogeography that uses patterns of evolution within species that are discernible by analysis of molecular data. Such studies are becoming increasingly common for Alabama's turtles, and the intraspecific lineages generated by such studies likely allow the discovery of new lineages that were not evident from analysis of traditional morphological data. This creates an exciting environment for taxonomists because so many new lineages may be species available for discovery, and such discoveries tell us important stories about how Alabama's rich biodiversity was generated (e.g., Soltis et al. 2006). It also creates a stressful environment for authors of field guides, such as this one, because of the likelihood that the guide will be obsolete before it is published. Nevertheless, we attempt to describe all lineages supported by character data of any kind, and, therefore, that might indicate speciation events awaiting taxonomic recognition.

Turtles—Testudines

Turtles, as a group, are our oldest living amniotes, having been on the earth for at least 220 million years and having diverged from the common ancestor to all other lineages 280 million years ago. While varying considerably in size, shape, and color, all turtles are easily recognizable as such. Each species has a shell that consists of an upper portion, the carapace, which joins laterally with a lower portion, the plastron. A close inspection of the shell of most turtles will reveal an inner layer of bony plates created by modified ribs, vertebrae, and elements of the sternum, and an outer layer of horny epidermal scutes. These structures characterize all turtles possessing a hard shell. The shell becomes modified in two radiations of turtles that have soft shells. In these forms, the bony elements are greatly reduced, and the outer layer of the shell is composed of leathery skin rather than hard epidermal scutes.

Besides the shell, turtles have a number of other unique features. Primary among these is the use of specialized neck muscles to ventilate the lungs. The rib cage in most amniotes is designed to expand and contract, creating a pump mechanism to draw air into the lungs and then to force it out of the lungs. Because the ribs of turtles are solidly fused to the shell, neck muscles take over the role of lung ventilation by creating a bellows apparatus at the base of the neck. Associated with this change in anatomical structure, the shoulder girdle of turtles has moved from a position outside the rib cage, where it is positioned in all land vertebrates except turtles, to a position inside the rib cage. The sum total of changes that characterize living turtles has created a radiation that is immediately identifiable to all human cultures.

The life histories of turtles vary in details but are similar in basic aspects. All species are oviparous, laying eggs in leaf litter or, more frequently, in a cavity that the female digs in loose soil. Fertilization is internal, accomplished by means of a penis that is extruded from the vent of the male at the time of copulation and inserted into the cloaca of the female. Courtship activities prior to copulation vary, and the details of this behavior are known for relatively few species. In general, these courtship behaviors involve displays given by the male turtle to a female. Although not yet documented by careful study, females clearly use these courtship behaviors to select from among several competing males. In River Cooters (*Pseudemys*) and Painted Turtles (*Chrysemys*), the male gives a tactile cue by stroking the head of the female with elongate claws on his front limbs while swimming in front of her. In Gopher Tortoises, courtship is accompanied by visual cues (head-bobbing on the part of the male), tactile cues (biting of the female's head and anterior carapace by the male), and chemical cues (rubbing of exudates from subdentary glands by males and females onto their own and their partner's forelimbs). Occasionally, male Gopher Tortoises may pound the side of the female's shell with the front end of his plastron. Male Eastern Box Turtles bite and nip at the front of a female's carapace during courtship. Males of some Map Turtles (*Graptemys*) vibrate their head against the top and sides of a female's head; in other male *Graptemys*, claws are used in a manner similar to males of *Pseudemys*.

Because courtship activities of a male are designed to gain the attention of a female, female turtles appear to choose from available male turtles by maintaining or breaking off that attention. For example, in Slider Turtles (*Trachemys*), females frequently avoid interactions with courting males by diving underwater. If a male is selected, the female signals her choice by rapidly opening and closing the nictitating membrane, an opaque eyelid-like structure found in diving reptiles, which gives a white flash when closed. In Gopher Tortoises, females may ram courting males or may break off courtship by entering a burrow. If she selects a mate, then a female Gopher Tortoise signals this selection by walking backward while turning her back to the male.

Once selected by a female mate, the male turtle typically mounts the female from the rear, twisting his tail beneath hers. In several species of turtles, the male's plastron has a concavity that facilitates assumption of the copulatory position. This plastral concavity is especially pronounced in male Box Turtles, but is noticeable also in male

Gopher Tortoises and Eastern Mud Turtles. Male turtles may hold on to the female's shell with all four feet, with just the hind feet, or with the feet and tail. The tails of male mud and musk turtles are adapted to hook under the female's carapace during copulation, locking male and female turtles together. In back of the knee of the hind leg of males of these genera are opposable patches of horny skin in which the tail of the female is grasped and pulled to one side.

In all species of Alabama turtles, a nest cavity is excavated by the female using her hind feet. In some cases the female expels fluid from the bladder during the process; this softens the soil and makes digging easier. Some nests are never completed, either because the soil is not soft enough for digging or because incomplete nests fool predators as to the location of real nests. For complete nests, a female deposits eggs and then covers them using only the hind feet and smoothing the top of the nest with her plastron. The female then leaves the nest, providing no further parental care.

The shell of turtle eggs is composed of protein with aggregations of calcium. Tortoises, soft-shelled turtles, and musk and mud turtles have hard shells because of increased calcium concentrations and complex organization of the calcium. The shells of these eggs are white in color and have hard surfaces, as in chicken eggs. The remainder of Alabama's turtle species produces eggs that are soft and leathery because the calcium is in low concentrations and is unorganized in structure. These shells often are tan in color.

Sexual dimorphism is evident in most species of turtles. The tail of a male is longer and thicker than that of a female. The vent of a male is usually well beyond the edge of the carapace when the tail is extended, whereas in a female the vent is under the carapace. In most turtle species, an adult female attains a greater size than an adult male. Size difference is especially pronounced in some species of Map Turtles in which the female may be 2.5 times larger than the male. In these species the females not only exceed the males in size, but, in most females, develop greatly enlarged heads, a change that does not occur in the males. However, there are unusual cases in Alabama's turtle fauna in which the male grows to a larger size than the female (Snapping Turtles [*Chelydra* and *Macrochelys*]), or are approximately equal in size (Gopher Tortoise).

Sexual dichromatism also can be pronounced in some turtles. In soft-shelled turtles, the juvenile carapace pattern is often pronounced and includes large dark spots or dark rings along the edge of the back half of the carapace. In some species, males retain this juvenile coloration while females become uniform in color. In Red-eared Sliders, the reddish bar on the side of the head is retained well into adulthood in most females, but becomes obscured by melanism in most adult males.

For their size, all turtle species exhibit unusual longevity. Although predation on eggs and juveniles can be high, adults have life spans of sixty to eighty years, during which females apparently never lose fertility. Because adults live so long and individuals within populations have such great opportunities to develop long-term behavioral ties, social organization can be as complex as it is in some mammals. Male Gopher Tortoises, for example, remember the position of, on average, three females each breeding season and visit them on a more or less daily basis. Because females shift positions in the landscape from year to year, males likely have exceptional spatial memory in order to track potential mates. In addition to memory, male Gopher Tortoises are known to use chemicals to distinguish known neighbors from novel intruding males; female Gopher Tortoises use chemical cues to distinguish attractive from unattractive male mates.

Turtles display a wide variety of foraging modes. Like modern birds, turtles lack teeth and process food by slicing it with scissors-like jaws or by gulping the diet item. Several species are herbivorous, using the jaws to slice vegetation, which may be held in the esophagus prior to entering the stomach. Herbivorous turtles also have an enlarged caecum that holds bacteria capable of digesting cellulose. A few turtles have a generalized diet that includes both plant and animal material, but most are predators that consume a relatively restricted diet. Some of these diets are surprisingly specialized on unusual prey. Several members of the genus *Graptemys* specialize on mollusks, which are crushed by broad shelves of bone inside their massive heads. Members of the Leatherback Sea Turtle (*Dermochelys*) eat jellyfish whose stinging cells may be neutralized by large papillae that cover the esophagus.

Turtles have been used as a food source for humans essentially since the derivation of the human species. Because turtle populations reach high densities, this source of meat has been substantial, especially in the southeastern United States, where turtle species richness is unusually high. Not only do turtles constitute a valuable food resource, but they are the center of folklore for native human cultures of the United States, and provide a source of enjoyment for a growing number of people in urban areas who like to keep them as pets or merely observe them in their natural habitats. Box Turtle shells are still used as rattles by Native Americans in their dance rituals and, according to some Native American folklore, the earth was created on a turtle's back who continues to carry it (Langford 1987).

The world's turtle fauna consists of about 340 species and is divisible into two major groups, Pleurodira and Cryptodira. The living pleurodirans, or side-necked turtles, comprise three families with nineteen genera that occur in South America, Africa, Madagascar, New Guinea, and Australia. However, this radiation split from other turtles 210 million years ago (Near et al. 2005), when the position of the current continents was far different than it is today. In fact, at this time, Alabama was near the equator and was attached to northern South America (Lacefield 2013). For this reason, Alabama's fossil turtle fauna includes pleurodirans that were present in the state as late as 84 million years ago in the form of two genera, *Chedighaii* and *Bothremys,* of freshwater turtles (Gaffney et al. 2009). These genera are relatives of the modern river turtles (*Podocnemis*) of South America, and likely diversified in equatorial parts of North and South America 120 million years ago (Near et al. 2005).

Cryptodira, with eleven families and seventy-three genera worldwide, is by far the larger group of living turtles and is represented on every continent except Antarctica. All living North American turtles are cryptodirans and many fossil forms belonging to this lineage are known. Alabama's turtle fauna is exceptionally rich relative to those of all other areas of comparable size in the world. It consists of thirty-two living species, within seven families and eighteen genera.

Turtles have been exploited by humans for centuries, a pressure that has imperiled over half of all species (Turtle Conservation Coalition 2011). In Southeast Asia, exploitation is so intense that the entire turtle fauna of this region is nearing extinction, creating a market for exploitation of turtles from elsewhere, including Alabama. If the turtle radiation is to remain as part of the world's fauna, then expanded conservation efforts must be mounted now to prevent further decline. In response to these conservation needs, the Alabama Department of Conservation, in 2012, amended state regulations to ban the take of native turtles in public or private waters for commercial purposes. This action protects the state's entire turtle fauna from export markets. Turtles can be raised in the state for commercial use, but this must be done by licensed farmers or dealers who cannot use wild-caught animals. Additionally, these facilities are prohibited from importing non-indigenous species to be propagated within the state. Turtles may be exploited within the state for use as food or pets, but only for those species that are not protected by federal or state laws. Two unprotected turtles per day may be collected for personal use, but these animals cannot be sold or traded for anything of monetary value. Additionally, unprotected turtle species may be removed from ponds on private property if populations reach nuisance levels, but these may not be sold or traded for anything of monetary value. These regulations place Alabama at the forefront of turtle conservation in the United States. However, Alabama's turtle populations remain susceptible to extirpation when mortality of adult females is increased (Congdon et al. 1987). Therefore, protection of riparian zones where females lay eggs is crucial to managing populations of these creatures.

Key to the Families of Turtles of Alabama

The conformation of the epidermal scutes of Alabama's turtles with hard shells is an important feature that allows identification of species and subspecies. These scutes are arranged in distinct rows, each of which is named based on the position of that row. These rows will be referred to in the keys that follow.

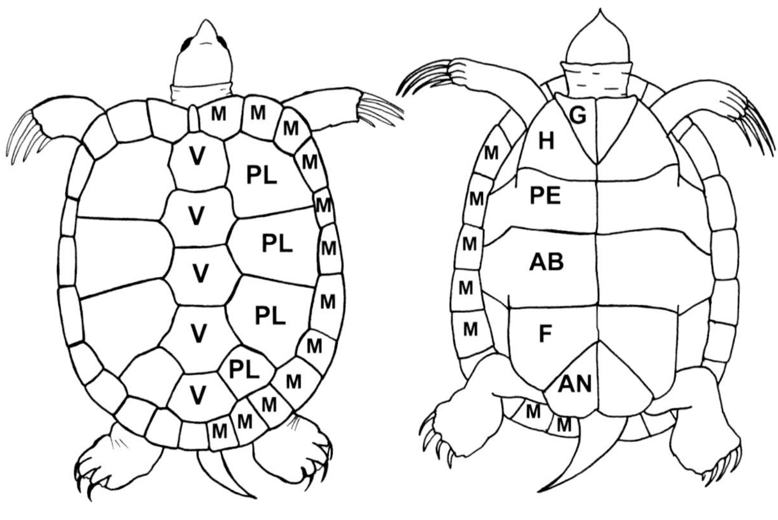

From left to right:

Epidermal scutes of turtle carapace:
V = Vertebral
PL = Pleural
M = Marginal

Epidermal scutes of turtle plastron:
M = Marginal
G = Gular
H = Humeral
PE = Pectoral
AB = Abdominal
F = Femoral
AN = Anal

1a Limbs modified into flippers. Go to **2**.
1b Limbs paddle-shaped or rounded. Go to **3**.

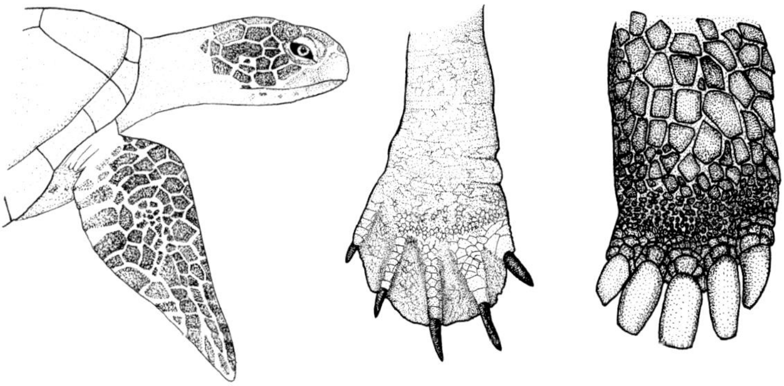

From left to right:

Forelimb flipper of Green Sea Turtle

Paddle-shaped forelimb of Softshell Turtle

Rounded forelimb of Gopher Tortoise

2a Shell without scutes; carapace with seven longitudinal ridges.
Family **Dermochelyidae—Leatherback Sea Turtles . . . . page 31.**

2b Shell with scutes; carapace, if keeled, having no more than four keels.
Family **Cheloniidae—Sea Turtles . . . . page 37.**

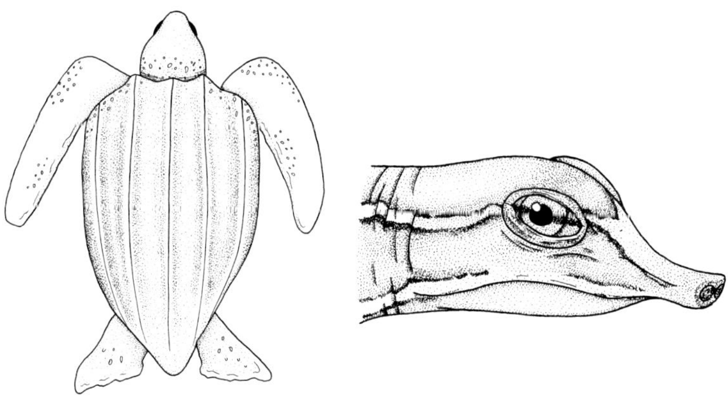

From left to right:

Dorsal view showing longitudinal ridges of Leatherback Sea Turtle

Head of Softshell Turtle showing tubular proboscis

3a Shell covered with leathery skin; forelimbs scale-less and shaped like fleshy paddles; snout ending in a tubular proboscis.
Family **Trionychidae—Soft-shelled Turtles . . . . page 61.**

3b Shell covered by hard scutes; forelimbs scaled; snout not ending in a tubular proboscis. Go to **4**.

4a Terrestrial; forelimbs shovel-like for digging; hind feet elephantine.
Family **Testudinidae—Tortoises . . . . page 83.**

4b Aquatic (except *Terrapene*); forelimbs webbed and paddle shaped; hind feet paddle shaped, adapted for swimming. Go to **5.**

5a Plastron small; tail long, about as long as carapace.
Family **Chelydridae—Snapping Turtles . . . . page 93.**

5b Plastron large, tail short, less than half the length of carapace. Go to **6.**

6a Plastron with ten or eleven scutes; pectorals of plastron not in contact with marginals.
Family **Kinosternidae—Mud and Musk Turtles . . . . page 105.**

6b Plastron with twelve scutes; pectorals touching marginals.
Family **Emydidae—River, Pond, and Box Turtles . . . . page 141.**

From left to right:

Small plastron of Eastern Snapping Turtle with long tail

Large plastron of Mud Turtle with short tail

Hinged plastron of Box Turtle with twelve scutes

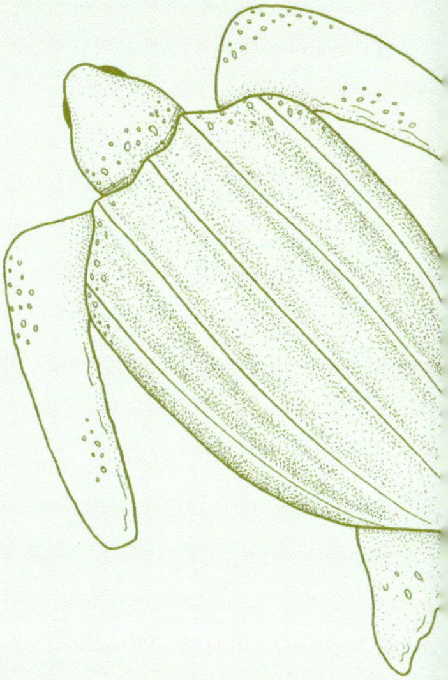

Leatherback Sea Turtles

Family Dermochelyidae

This family contains a single living genus, *Dermochelys* Blainville. The largest of all living turtles, these are called Leatherback Sea Turtles because, rather than having a hard shell of fused vertebrae and ribs, the shell is dominated by soft dermal layers in which small bones are embedded. Its range includes tropical seas in both hemispheres, but visits by Leatherbacks to the temperate zone are recorded regularly. This family is the sister group to the family Toxochelidae, an entirely fossil radiation of sea turtles that are common in Alabama Cretaceous deposits. Among living taxa, it is sister to the family Cheloniidae, from which it may have diverged 150 million years ago (E. S. Gaffney, pers. comm.).

Leatherback Sea Turtles

Genus *Dermochelys* (Blainville, 1816)

This sole genus in the family contains the only living species. Nine fossil genera, including the genus *Psephophorus* known from the Pliocene of Central Florida (Dodd and Morgan 1992), are related to *Dermochelys*. Therefore, this is an ancient lineage represented by a single survivor that is specialized for living in open ocean waters where it dives deep and consumes jellyfishes. Molecular data suggest that populations of this species evolved recently (less than 1 million years ago), likely originating in the Indo-Pacific, and expanding to occupy the Atlantic Ocean and Gulf of Mexico (Dutton et al. 1999).

Atlantic Leatherback

Dermochelys coriacea (Vandelli, 1761)

Description This species is an enormous salt-water turtle that attains a maximum carapace length of 96 inches (2,438 mm), making it the world's largest living turtle. The species is unmistakable on the basis of the extremely elongate wing-like forelimbs and smooth skin of the carapace and plastron, which have no scutes. The carapace has seven longitudinal ridges and there are five ridges on the plastron. In color this species can be black, gray, or dark brown with patches of white. Young are covered with numerous small scales that are more vividly marked than in adults.

Alabama Distribution This species has been documented for Alabama based on rare instances of stranded adults or observations of foraging adults. Nesting is possible for the state based on a confirmed nesting event at Gulf Island National Seashore (Florida) in 2000 (US Department of Interior, revised 2008).

Habits Leatherbacks feed on jellyfish, generally in areas of open oceans and in cooler waters. During dives they are capable of descending to 3,900 feet (1,186 m) and may remain submerged for

Right: Hatchling Leatherback Sea Turtle, Chatham County, GA

Previous: Adult Leatherback Sea Turtle, Chatham County, GA

eighty-six minutes (Lopez-Mendilaharsu et al. 2008). Their travels between nesting grounds and feeding areas may take them 8,700 miles (14,000 km) in a year (Hughes et al. 1998). Leatherback turtles in the Atlantic nest mostly in Puerto Rico, the island of Culebra, and the Virgin Islands. But about fifty nests occur in Florida each year, mostly on the eastern side of the state. Like all sea turtles, females crawl high onto the beach to dig nests. But Leatherback Sea Turtles tend to select beaches with fine sands and shallow slopes. Their crawl marks are distinctive because of the extreme width (6 feet [1.8 m]) between the impressions left by the forelimbs. The marks are symmetric (right and left limbs move simultaneously) with a tail drag mark down the center. Individuals nest between February and July and females may lay up to nine clutches of about 100 eggs each. The sex of each offspring is determined by the temperature of the nest, with warmer nests producing a greater proportion of females (Standora and Spotila 1985). Like other sea turtles, females nest every two to three years.

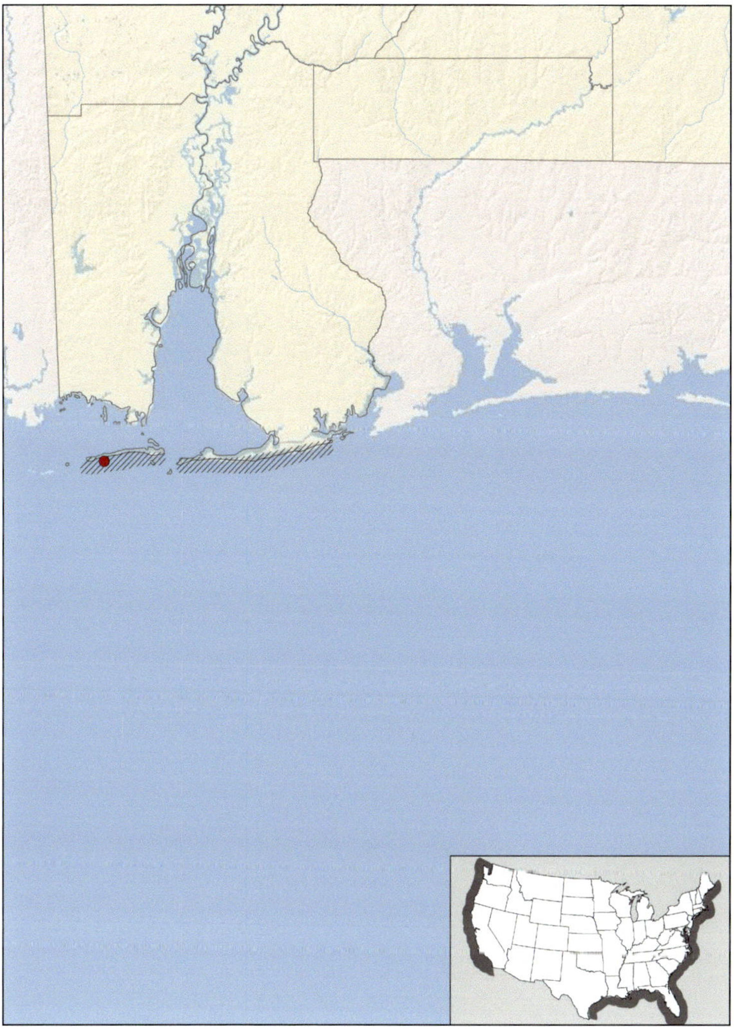

Alabama and continental US distribution of the Leatherback Sea Turtle

Conservation and Management Because of worldwide reduction in populations of all sea turtles, the species is listed as endangered by the ESA and critically endangered on the IUCN Red List and in Appendix 1 on CITES. The Atlantic Leatherback is also ranked Priority 1 (Highest Conservation Concern) by ADCNR (Mirarchi et al. 2004) and, therefore, is protected by the state of Alabama. The flesh of this species is considered inedible, but the eggs are eaten regularly in some parts of the world. Additionally, adults are frequently drowned in gill nets and longline operations of commercial fishermen. Use of turtle excluder devices likely reduces drowning in shrimp nets. A recovery plan was approved by USFWS and NOAA National Marine Fisheries Service in 1992, and this plan indicates recovery can be achieved if nesting populations in the Atlantic increase over twenty-five years and 75 percent of nesting beaches are in public protection. However, because of the limited likelihood that this species will nest in Alabama, the state will not be crucial in the conservation of this species.

Taxonomy No subspecies of this cosmopolitan marine turtle are currently recognized, although eleven mitochondrial haplotypes have been identified and divide the species into Indo-Pacific, Pacific, and Atlantic populations (Dutton et al. 1999). Previous publications have placed this species in the genera *Chelonia, Dermochelis, Dermatochelys, Testudo,* and *Sphargis.*

Sea Turtles

Family Cheloniidae

The family Cheloniidae is represented in warm seas throughout the world. Turtles in this family are large in size, have hard, bony shells, and are designed for life at sea. The front limbs are modified into wing-like flippers that are used to propel the animals through water, with the paddle-shaped hind limbs being used as rudders. These turtles can dive to depths that exceed 3,300 feet (1,000 m) and may remain underwater for 1.5 hours when feeding (Lopez-Mendilaharsu et al. 2008) and for nearly seven hours when resting (Hochscheid et al. 2005). The oxygen debt developed during a dive is repaid rapidly when an individual surfaces to breathe and so sea turtles spend about 90 percent of their time underwater, a percentage that far exceeds that of diving birds and mammals (Hochscheid et al. 1999). These dives allow sea turtles to forage and rest underwater. Sea turtles migrate from nesting beaches to foraging ranges during a season of activity, movements that may exceed 1,860 miles (3,000 km). When on land, seas turtles move slowly and clumsily. Because they live in salt water, a salt gland, located in the corner of each eye, exudes excess salt, often creating a stained streak below the eye when viewed on land.

This family includes five genera (six species), *Caretta, Chelonia, Eretmochelys, Lepidochelys,* and *Natator,* with all but *Natator* occurring along the Alabama coast. Only *Caretta* and *Lepidochelys,* however, habitually nest on our beaches. This family is sister to Osteophygidae (E. S. Gaffney, pers. comm.), an entirely fossil radiation of sea turtles. Among living families, it is sister to the family Dermochelyidae, having diverged from that family at least 50 million years ago (Near et al. 2005).

Key to the Genera of Cheloniidae of Alabama

1a Pleurals in five or more pairs, first pair smaller than rest. Go to **2.**
1b Pleurals in four pairs, all large. Go to **3.**

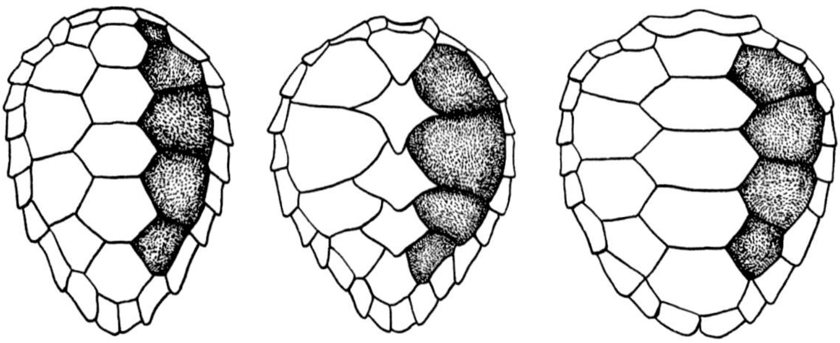

From left to right:

Carapace of Loggerhead Sea Turtle with pleurals in five pairs, first small

Carapace of Hawksbill Sea Turtle with overlapping pleurals in four pairs

Carapace of Green Sea Turtle with pleurals in four pairs, all large

2a Bridge with three enlarged scutes.
Genus ***Caretta*—Loggerhead Sea Turtles . . . . page 39.**
2b Bridge with four enlarged scutes.
Genus ***Lepidochelys*—Ridley Sea Turtles . . . . page 46.**

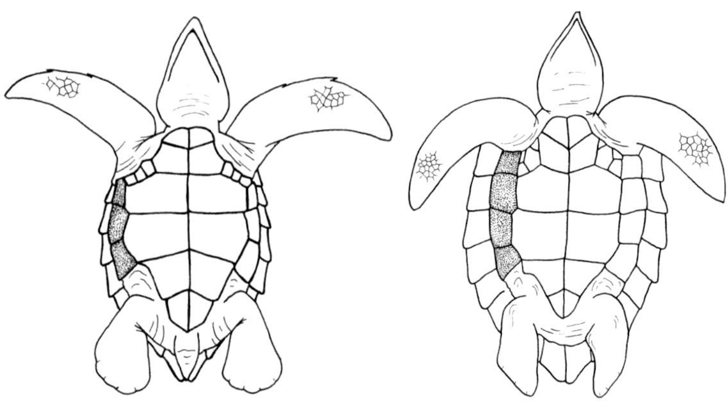

From left to right:

Plastron of Loggerhead Sea Turtle, bridge with three scutes

Plastron of Ridley Sea Turtle, bridge with four scutes

3a Scutes of carapace overlapping.
Genus ***Eretmochelys*—Hawksbill Sea Turtles . . . . page 52.**
3b Scutes of carapace not overlapping.
Genus ***Chelonia*—Green Sea Turtles . . . . page 56.**

Loggerhead Sea Turtles

Genus *Caretta* (Rafinesque, 1814)

This genus, containing only one living species, is widely distributed in warm oceans of both hemispheres. It differs from all other sea turtles in having a hard shell, with five pairs of pleural scutes on the carapace and a bridge composed of three scutes. Along with *Dermochelys*, *Caretta* tends to inhabit deeper waters where it dives to greater depths than those sea turtle genera that live near shore (Hochscheid et al. 1999). The broad distribution of *Caretta* makes it the most frequently encountered sea turtle on nesting beaches worldwide. Its sister taxon is the genus *Lepidochelys*, both of which are predatory.

Loggerhead Sea Turtle

Caretta caretta (Linnaeus, 1758)

Description Loggerhead Sea Turtles can be huge, salt-water creatures attaining a maximum carapace length of about 84 inches (2,130 mm). The feet are modified into wing-like flippers and the head is large, with two pairs of prefrontal scales. The carapace is dominated by five pairs of pleural scutes and has three keels, one down the vertebral scutes and a pair down the center of the pleurals on each side. Scutes of the carapace abut rather than overlap each other, and the scutes that form the mid-dorsal keel become less pronounced in older individuals. The front of the carapace often has a concavity. The plastron has two keels. Three enlarged scutes cover the bridge, separating a reddish brown carapace from a white plastron in adults, and a light brown carapace from a white plastron in juveniles. The skin of adults and juveniles is brown or yellowish. Adult males have a longer and broader tail, longer claws, and a shorter plastron than adult females. Additionally, males have larger heads than females. This species differs from all other of Alabama's sea turtles in having five pleural scutes on the carapace and a bridge composed of three scutes.

Alabama Distribution Loggerhead Sea Turtles occupy coastal waters of Mobile and Baldwin Counties.

Left: Hatchling Loggerhead Sea Turtle, Chatham County, GA

Opposite: Adult female Loggerhead Sea Turtle, Australia

Habits The Loggerhead Sea Turtle is the most frequently encountered marine turtle along the Alabama coast, and it is the only one that nests on our beaches in substantial numbers. Nesting may occur anywhere from the Florida border to the end of the Fort Morgan Peninsula in Baldwin County and along the length of Dauphin Island in Mobile County.

Loggerheads are almost wholly carnivorous, feeding on a variety of crustaceans, mollusks, echinoderms, and fish, when foraging in shallow waters near shore, and on soft-bodied cnidarians when foraging in deep waters. It is reported to feed on Portuguese man-of-war, keeping its eyes closed to protect it from repeated stings. The massive head and broad jaws create great power for crushing mollusk shells and crustacean exoskeletons, and the finger-like papillae that line the esophagus are thought to filter items as they are swallowed. The species spends about 60 percent of its time underwater, and 90 percent of foraging dives are 130 feet (40 m) or less in depth (Polovina et al. 2004). While they are on the foraging grounds, females defend territories by aggressively driving off intrusion by other females (Schofield et al. 2007).

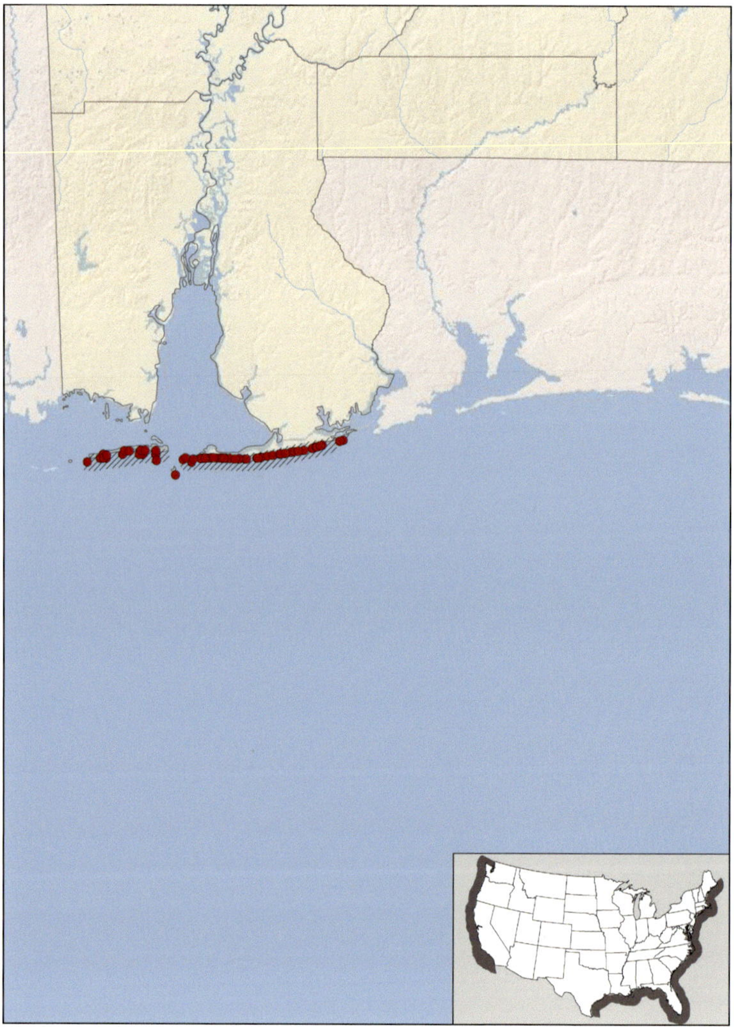

Alabama and continental US distribution of the Loggerhead Sea Turtle

Like other sea turtles, Loggerheads migrate between foraging and nesting sites, traveling as much as 1,240 miles (2,000 km) from warm southern waters in winter to cool northern waters in summer (Polovina et al. 2004). Individuals captured on Alabama beaches and monitored with telemetry equipment typically remain along the edge of the continental shelf of the Panhandle of Florida and Alabama, wandering as far south as the coast of Cuba. This species is distributed farther north than Alabama's other sea turtles and occupies waters where cold spells may stun them into inactivity. In cold waters during winter months, resting dives of up to seven hours may occur, a pattern Hochscheid et al. (2005) described as hibernation.

Mating occurs in early spring, during migratory movements from foraging areas, and involves induced ovulation, a physiological mechanism in which the act of mating provides the final stimulation of the female hormone cycle to cause ovulation to occur (Manire et al. 2008). This feature assures that fresh sperm are available for fertilization of the eggs. However, females also store sperm, a feature that allows up to five males to father offspring in a single clutch of eggs (Bollmer et al. 1999; Zbinden et al. 2007). Nesting occurs from May through August, with a peak in mid-June. A gravid female intent on nesting crawls from the water, usually shortly after dark, and ascends to a point somewhere between the high-tide mark and the dune front. Her tracks are distinctive in being about 3 feet (0.9 m) wide, shaped like commas, and alternating between use of right and left flippers (indentations in sand are offset; US Department of Interior 2008). Females prefer beaches with steep slopes that are free of debris. At such places females crawl high above the tide seeking relatively dry sands. About half of emerging females return to the water without depositing eggs. These false crawls may indicate exploration by or disturbance of gravid females that will return to lay eggs later. Nesting females use all four limbs to scoop out a shallow depression and then, using the hind flippers only, they dig a nest chamber that ranges in depth from 18 to 26 inches (46 to 66 cm). The eggs, roughly spherical in shape, are deposited in clutches that average 110 eggs; up to five clutches may be deposited during a single nesting season (US Department of Interior 2008). Each nest is covered with sand and packed. Before leaving the nest, the female makes sweeping movements with her front and hind limbs, scattering the sand. This process leaves substrate that is mounded above the nest and disturbances at the nest site that are wider than the body of the female. These characteristics (mounding and disturbance width) distinguish nest sites from false crawls. Because reproductive effort is so costly, females take two to three years to store enough energy to produce the next clutches of eggs.

Predation on the incubating eggs is heavy. Raccoons, foxes, feral dogs, and sand crabs are the most serious predators. In some places, hogs destroy nests as well. Incubation to hatching takes an average of sixty-five days, and, assuming a clutch hatches successfully, the baby turtles wiggle and thrash about in the cavity until they break through to the surface. The sex of the offspring is determined by the temperature of the nest, with warmer nests producing a greater proportion of females (Standora and Spotila 1985). Emergence usually occurs at night, and those few that break through during the day are often killed by the heat of the sun or are consumed by predators. The hatchlings from a given nest usually appear at the surface in groups, and several days may elapse before the entire brood has left the nest.

Conservation and Management The Loggerhead Sea Turtle is ranked Priority 1 (Highest Conservation Concern) by ADCNR (Mirarchi et al. 2004) and, therefore, is protected by the state of Alabama. The flesh of the Loggerhead Sea Turtle is not highly regarded as food, and so adults of this species generally are not killed for their meat. However, eggs are consumed by humans and, worldwide, nest robbing is a serious problem with maintaining populations of Loggerheads. However, this is not a great problem anymore in Alabama. Chronic loss of adult Loggerhead Sea Turtles occurs when they drown after being captured in shrimp nets. Turtle excluder devices are required by NOAA for all shrimp nets in the Gulf of Mexico, and these reduce the number of turtles that are killed relative to mortality without the devices. In the twenty years (1987 to 2007) preceding the Deepwater Horizon oil spill of April 2010 in the Gulf, an average of five stranded Loggerhead Sea Turtles was reported for Alabama each year. During 2010 and 2011, this increased to forty-eight strandings per year (http://www.nmfs.noaa.gov/pr/species/turtles/gulfofmexico.htm).

Since 2001, Share the Beach, a sea turtle volunteer program, has monitored Alabama beaches for nesting turtles. The goal of the monitoring is to discover all nests along Alabama's coasts, to protect them from mammalian and avian predators, and to count hatchlings and then assist these hatchlings to the sea. Monitoring takes place primarily from May through October from the Florida border to the tip of the Fort Morgan Peninsula and along Dauphin Island. Volunteers are trained to specifications established by USFWS. Nests are marked and, starting fifty-five days after deposition, monitored each night until hatchlings start to emerge. The hatchlings are assisted to the tide line by a trench that is dug from the nest to the water. From 2006 to 2013, an average of 4,200 hatchling Loggerheads per year were produced on Alabama beaches from an average of seventy-five nests each year (http://www.alabamaseaturtles.com/nesting-season-statistics/, accessed March 8, 2014).

Loggerheads are listed as threatened under the ESA and endangered on the IUCN Red List and in Appendix 1 by CITES. NOAA also regulates fishing intensity and can ban fishing from areas that might need protection for Loggerhead Sea Turtles. But because this is the most common sea turtle, management efforts are not intense and populations continue to decline. A plan created in 1991 and signed in 2008 lists production of 300 nests per year as constituting Alabama's contribution to recovery of the North Gulf of Mexico Management Unit of Loggerheads; this increase must result from increased numbers of nesting females. Considerable effort will be required to achieve this goal. The state wildlife action plan lists expanding public holdings along Alabama's beaches and restoring native dune vegetation as key goals for maintaining nesting habitat for Loggerhead Sea Turtles (Alabama Department of Conservation and Natural Resources, Division of Wildlife and Freshwater Fisheries 2005).

Taxonomy We consider this to be one polymorphic species with no distinctive subspecific variation. Others consider it to consist of two subspecies, one in the Atlantic Ocean and one in the Pacific. It is known to hybridize with *Chelonia* and *Eretmochelys* (James et al. 2004; Lara-Ruiz et al. 2006). Previous literature has placed this species in the genera *Caouana, Cephalochelys, Chelonia, Testudo, Thalassochelys,* and *Thalassiochelys.*

Ridley Sea Turtles

Genus *Lepidochelys* (Fitzinger, 1843)

This genus contains two species and has a wide distribution in the warm marine waters of both Eastern and Western Hemispheres. The common name for these animals is Ridley Sea Turtle, a name of unknown origin. One of the two species inhabits the Gulf of Mexico and visits Alabama beaches. The genus tends to remain near shore and is famous for the mass arrival of gravid females to restricted nesting beaches over short time periods associated with the new moon of April through August. The sister taxon of ridleys is the genus *Caretta*, both of which are predatory sea turtles.

Kemp's Ridley

Lepidochelys kempii (Garman, 1880)

Description This is the smallest of the Atlantic sea turtles, attaining a maximum carapace length of around 29 inches (750 mm). The feet are modified into flippers and the head is large, containing two pairs of prefrontals. A short, wide, nearly circular carapace is dominated by five pairs of pleural scutes, the first touching the nuchal scute. Scales of the carapace abut rather than overlap each other, and there is a keel on the carapace that is created by the vertebral scutes. The bridge is composed of four enlarged scutes, and separates the gray carapace from the white or cream plastron. Young are darker than adults and have four tuberculate ridges on the carapace and three on the plastron. This species differs from all of Alabama's other sea turtles in having five pleural scutes and a bridge composed of four scutes.

Alabama Distribution Kemp's Ridley is found in the Gulf of Mexico and has been confirmed in Alabama by nests discovered on beaches near Fort Morgan, Laguna Key, Alabama Point (Gulf State Park), and Bon Secour National Wildlife Refuge. Stranding records indicate that this is a common species along Alabama's inshore waters, but nesting largely occurs elsewhere.

Right: Hatchling Kemp's Ridley Sea Turtle, Chatham County, GA

Previous: Adult Kemp's Ridley Sea Turtle, Chatham County, GA

Habits Kemp's Ridley was at one time called by some the "bastard turtle" because it was thought to result from hybridization of a Loggerhead and a Green Sea Turtle. Although there is no evidence that such hybridization maintains this species, the mitochondrion of Kemp's Ridley carries evidence of a past hybridization with the Loggerhead Sea Turtle (James et al. 2004; Lara-Ruiz et al. 2006). The species is capable of long migratory movements encompassing at least 1,615 miles (2600 km) and, during periods of activity, spends 89 percent of the time in dives that may be as deep as 460 feet (140 m; Renaud 1995).

The primary nesting beaches for this species are along the shores of extreme northern Tamaulipas, Mexico, and the Padre Island National Seashore of extreme southern Texas. However, isolated nesting occurs along other beaches of the Gulf of Mexico, including Alabama. Persistent but rare nesting of this species occurs in the state. Females crawl to the vegetation line along seashores to dig nests where they deposit about 100 eggs. The nest is then covered and the female returns to the sea, but she may return during a nesting season to deposit

Alabama and continental US distribution of the Kemp's Ridley Sea Turtle

up to three additional clutches of eggs. The tracks created by a nesting female are distinctive, exhibiting alternate (offset) comma-shaped marks of the front flippers that are about thirty inches apart (US Department of Interior 2008). Additionally, females of this species are unusual in generally arriving to nest during daylight hours.

Ridleys consume crustaceans, mollusks, anthozoans, and seaweeds that are crushed by jaws designed for grinding food. Juveniles migrate throughout the Gulf of Mexico and eastern Atlantic where they are found primarily in beds of sargassum, common seaweed of the Atlantic Ocean. Most individuals along Alabama's coast are juveniles that grow to adulthood and then migrate to the primary nesting beaches in northern Mexico and southern Texas.

Conservation and Management Kemp's Ridley is ranked Priority 1 (Highest Conservation Concern) by ADCNR (Mirarchi et al. 2004) and, therefore, the species is protected by the state of Alabama. The flesh of Kemp's Ridley is considered inferior to that of the Green Sea Turtle, so predation by humans is reduced relative to other sea turtles. However, this has not prevented the Kemp's Ridley from becoming the most endangered of the world's sea turtles. The species is listed as endangered under the ESA and the IUCN Red List, and appears in Appendix 1 of CITES. The primary causes of imperilment are collection of eggs for human consumption, sacrifice of adults for boot leather, and drowning of juveniles and adults in shrimp nets. To these chronic sources of mortality can be added acute bouts of increased mortality associated with unusually cold winters, which can increase death by drowning of immobilized individuals. The Gulf oil spill of April 2010 also represents an acute mortality event. During the twenty years before this spill, two to three strandings of Kemp's Ridleys were reported annually for Alabama beaches. During the year after the spill, sixty strandings were reported (http://www.nmfs.noaa.gov/pr/species/turtles/gulfofmexico.htm). Of 600 instances of sea turtle deaths during the first six months after that spill, 79 percent were of Kemp's Ridleys.

Management activities that have centered on Kemp's Ridleys include a major collaborative project between Mexico and the United States to collect eggs from the Rancho Nuevo nesting site in Tamaulipas, Mexico, to incubate the eggs in sand from Padre Island National Seashore in south Texas, and then to release the hatchlings at Padre Island. This process has been used to attempt to create a second primary nesting site. Hatchlings at this site are allowed to enter the Gulf of Mexico at Padre Island, where they are expected to imprint on the sand and water, and then are immediately recaptured and reared for nine to eleven months in captivity. This headstart activity is designed to raise individuals to a size that is too large for major predators of hatchling sea turtles. In collaboration with some seafood companies, conservation efforts at Rancho Nueva have been implemented to improve protection of that nesting beach. Nightly patrols during the nesting season have reduced poaching of adults and eggs. The conservation goal is to increase the population of Kemp's Ridleys to a level of 10,000 nesting females each year. From a low of about 700 nesting females in 1986, conservation activities had increased population size to an estimated 5,000 nesting females in 2010, data that suggest conservation efforts for this species will reach the goal. From this, we can expect increased nesting of the species on Alabama's beaches. As long as the Share the Beach program is active, most nests in Alabama should be discovered and protected. The goals of expanding public holdings along Alabama's beaches and restoring native dune vegetation listed in the SWAP (ADCNR-DWFF 2015) further solidify nesting habitat for Kemp's Ridleys in Alabama.

Taxonomy This species is the sister taxon of the Olive Ridley, *Lepidochelys olivacea*, of the Pacific Ocean. No subspecific variation is recognized. Previous authors have placed this species in the genera *Chelonia*, *Testudo*, and *Thalassochelys*.

Hawksbill Sea Turtles

Genus *Eretmochelys* (Fitzinger, 1843)

Eretmochelys is a monotypic genus that is widely distributed in warm marine waters of both the Eastern and Western Hemispheres. This species is most common in shallow tropical waters where it tends to have shallow dives (maximum of 79 feet [24 m]). These turtles are termed Hawksbill Sea Turtles in reference to the shape of the jaws, which are similar in appearance to those of raptors. It is the sister genus to *Chelonia* + *Natator*, grouping with sea turtles that appear to be united by a grazing diet, among other features (Lutz and Musick 1997). The carapace is exceptionally colorful and once was harvested by humans as tortoise shell used for lampshades, combs, and guitar picks.

Eretmochelys imbricata (Linnaeus, 1766)

Taxonomy Two subspecies of this marine turtle are recognized, one of which occurs in the Gulf of Mexico and might be expected along our coast. This species is known to hybridize with the genera *Caretta* and *Chelonia* (James et al. 2004; Lara-Ruiz et al. 2006). Previous studies have placed this species in the genera *Caretta, Chelone, Chelonia, Onychochelys,* and *Testudo.*

Atlantic Hawksbill

Eretmochelys imbricata imbricata (Linnaeus, 1766)

Description Atlantic Hawksbill Turtles are marine organisms that attain a maximum carapace length of around 45 inches (1,140 mm). As with all sea turtles, the feet of Hawksbill Sea Turtles are modified into wing-like flippers. Two pairs of prefrontals are found on the head of this species. The carapace is keeled (vertebral scutes), heart-shaped or shield-shaped, and it contains scutes that overlap each other (except in very old individuals). Four pairs of pleural scutes dominate the carapace, and the first of these does not touch the nuchal. In color, the carapace has bright markings of reddish brown, accompanied by small markings of tans and yellows. The skin is predominantly brown above and yellow below. Juveniles have one keel on the carapace and two on the plastron. The bright colors of the carapace distinguish this species from all other sea turtles of Alabama.

Alabama Distribution This species is an occasional visitor to the Gulf of Mexico, but there have been no confirmed sightings or strandings of this species on Alabama beaches.

Right: Alabama and continental US distribution of the Hawksbill Sea Turtle

Previous: Adult Hawksbill Sea Turtle, Bora Bora, Pacific Islands

Habits Hawksbill Sea Turtles frequent coastal waters of the American tropics and the species occurs along the northern Gulf Coast only as an occasional migrant. This subspecies is most abundant in Caribbean waters where it inhabits coral reefs. The species spends 92 percent of its time underwater in resting and foraging dives. The elongate beak is designed to snip food from crevices in the coral reef. A variety of invertebrate prey is consumed, but sponges are a primary component of the diet. Females lay as many as 200 eggs in nests on beaches in the Caribbean. The temperature of the nest determines the sex of the offspring, with warm nests producing mostly females and cool nests producing mostly males (Janzen and Paukstis 1991).

Conservation and Management The Hawksbill Sea Turtle is ranked Priority 1 (Highest Conservation Concern) by ADCNR (Mirarchi et al. 2004) and, therefore, is protected by the state of Alabama. This species is perhaps the most persecuted of all the sea turtles. The quality of its meat is said to compare favorably with that of the Green Sea Turtle. Its eggs are used extensively for food. But it is exploited principally for its carapace, which is the source of tortoise shell. Persistent harvesting of adults for these resources has caused a severe decline in abundance of Hawksbills. For this reason the species is listed as threatened under the ESA and critically endangered on the IUCN Red List, and appears in Appendix 1 on CITES.

The closest sighting of Hawksbill Turtles to Alabama is near the head of DeSoto Canyon, located at the edge of the continental shelf south of Baldwin County. This area is near the Pinnacles, a region of deepwater reefs that may be a persistent attraction for Hawksbill Sea Turtles. Nevertheless, these turtles have never been observed to nest on Alabama's beaches. Turtle excluder devices on shrimp trawlers are known to reduce bycatch of these turtles, a key conservation tool for assuring their long-term viability. Protection of nesting beaches and headstarting are other conservation measures for this species elsewhere within its range. The Share the Beach program (see Conservation and Management under *Caretta caretta*) would mark and protect any nest that might appear during the nesting season of *C. caretta*. About twenty Atlantic Hawksbills were stranded along beaches of the Gulf Coast states within six months of the Gulf oil spill of April 2010, again suggesting that this species is persistent in waters off Alabama. All twenty survived to be released to the Gulf. A 1993 Recovery Plan for the species focuses conservation activities on nesting beaches of the US Virgin Islands and Puerto Rico, where recent nest surveys suggest populations are increasing. The goal of the Recovery Plan is to improve nest survival to 75 percent, to end trade in products, and to protect the foraging range of the species.

Green Sea Turtles

Genus *Chelonia* (Brongniart, 1800)

Chelonia contains the Green Sea Turtle, a largely herbivorous sea turtle that is distributed widely in warm oceans of both hemispheres. The genus gets its common name from the fact that fat deposits under the carapace are green in color. These turtles prefer shallow waters during most of their daily activities where they are commonly observed by tourist and fishing vessels. However, these animals also migrate great distances between nesting and foraging grounds. The genus contains only a single species. Its sister taxon is the genus *Natator* of the coast of Australia (Bowen and Karl 2007).

Green Sea Turtle

Chelonia mydas (Linnaeus, 1758)

Description Green Sea Turtles are large salt-water species attaining a maximum carapace length of at least 60 inches (1,525 mm). Their feet are modified into wing-like flippers, their heads contain two prefrontal scutes, and their shells are broad, low, flattened, and keelless in adults. Juveniles have one weak keel on the carapace (formed from the vertebral scutes) and two on the plastron. Four pairs of pleural scutes dominate the carapace, the first of which fails to touch the nuchal. Scutes of the carapace abut each other and do not overlap. In color, the carapace is light to dark brown, often with mottled blotches that may form radiating lines of darker color. The skin is brownish dorsally and light below. In juveniles the ends of the flippers are black, edged with white. Among Alabama's sea turtles, only *Eretmochelys imbricata* also has four pleural scutes, but that species has carapacial scutes that overlap each other; and these are extremely colorful compared to those of *Chelonia mydas*.

Alabama Distribution Although the usual range of this turtle in the Gulf of Mexico is well south of Alabama, an occasional straggler occurs along our coast, as indicated by two confirmed strandings between 1987 and 2007 and nesting in 2012.

Right: Hatchling Green Sea Turtle, Australia

Previous: Adult Green Sea Turtle, Australia

Habits In Atlantic waters, the Green Sea Turtle is principally an inhabitant of shallow, tropical shores. It is most common where vegetation, especially turtlegrass (*Thalassia testudinum*), is abundant. This species has foraging dives of up to 197 feet (60 m) in shallow waters, where it feeds on bottom organisms, but restricts dives to about 66 feet (20 m) in deep water, where it feeds on organisms suspended in the water column (Hawkes et al. 2006). Individuals may migrate up to 1,240 miles (2,000 km) between feeding and nesting sites (Troëng et al. 2005). Based on observations from Florida, nesting occurs from April to August and mating generally occurs off the nesting beach. The tracks of nesting females are distinctive in being about 4 feet (1.2m) wide, shaped like slash marks, and indicating simultaneous use of right and left flippers (indentations not offset; US Department of Interior 2008). Eggs number between 100 and 200 per clutch, and females can produce up to seven clutches during a nesting season. The sex of the offspring is determined by the temperature of the nest, with warmer nests producing a greater proportion of females (Standora and Spotila 1985). It takes a female two to four years to generate enough energy to produce the next clutch, but males attempt to mate every year.

Alabama and continental US distribution of the Green Sea Turtle

Conservation and Management The Green Sea Turtle is ranked Priority 1 (Highest Conservation Concern) by ADCNR (Mirarchi et al. 2004) and, therefore, the species is protected by the state of Alabama. Green Sea Turtles are prized for their flesh and eggs. In fact, meat from Green Sea Turtles is considered to be among the tastiest of vertebrate meats. Because adults aggregate at nesting beaches, overharvesting of both adults and eggs has caused populations to decline steadily in abundance. This has led to the listing of Green Sea Turtles as endangered under the ESA, endangered on the IUCN Red List, and in Appendix 1 by CITES. These actions are designed to restrict taking of adults and eggs.

The first possible nesting event for Green Sea Turtles on Alabama beaches was reported in 2012 by the Share the Beach program (see Conservation and Management under *Caretta caretta* for a description), which marked and protected the nest, as it does with all sea turtle nests. As long as this program continues, the species is adequately protected by efforts within Alabama. After the Gulf oil spill of April 2010, three stranded Green Sea Turtles appeared on Alabama beaches during 2011. Because only two had been observed in the twenty previous years, this was an unusual event. Of living Green Sea Turtles collected on beaches or in oil slicks, 86 percent survived long enough to be cleaned and released. Elsewhere, conservation efforts for this species in the United States are fairly intense, focused on nesting beaches of east central and southeast Florida. As indicated in a 1991 Recovery Plan for this species, conservation efforts focus on purchasing nesting beaches and improving hatchling survival with a goal of doubling the nesting population within the United States. The SWAP (Alabama Department of Conservation and Natural Resources, Division of Wildlife and Freshwater Fisheries 2005) lists similar actions for improving nesting habitat for Green Sea Turtles in Alabama. Additionally, this action plan lists enhancing sea grass beds as potential foraging habitat for Green Sea Turtles in Alabama's waters.

Taxonomy Although Carr (1952) recognized as many as eight statistically separable populations of *C. mydas*, we follow Iverson et al. (2008) in awaiting more definitive evidence of subspecific differentiation. However, some still distinguish Pacific from Atlantic populations. This species is known to hybridize with *Caretta* and *Eretmochelys* (James et al. 2004; Lara-Ruiz et al. 2006). Previous publications have placed this species in the genera *Caretta*, *Chelone*, *Euchelus*, *Mydas*, *Natator*, *Testudo*, and *Thallasiochelys*.

Soft-shelled Turtles

Family Trionychidae

This family occurs in North America, Africa, South Asia, and Indonesia. It contains about thirty species included within thirteen genera. All are highly aquatic with extensive webbing on all four limbs and a soft, pliable shell that is heavily flattened dorsoventrally. The family is most closely related to the family Carettochelyidae from which it diverged 155 million years ago (Near et al. 2005). Fossil materials belonging to this family are present in Alabama documenting that the family has occupied the state for at least the last 48 million years. Only a single genus occurs in North America.

North American Softshells

Genus *Apalone* (Rafinesque, 1832)

This genus occurs in North America and contains four species, all of which are found in Alabama. The genus diverged from its closest living relative, the genus *Rafetus* of the Middle East and China (Engstrom et al. 2004), about 100 million years ago (Near et al. 2005), suggesting that this divergence was associated with tectonic activity that separated North America from Europe and Asia. Softshells live in large rivers and streams, as well as ponds and lakes, and are designed to burrow in loose substrate, where they may remain submerged for long periods. North American Softshells prefer sand or muddy substrates and typically bury themselves where they can remain hidden but also extend their necks to the water surface to breathe. Buried softshells leave a circular impression at the substrate surface indicating the position of the turtle. The skin covering the plastron is highly vascularized and, along with the lining of the mouth and cloaca, is capable of exchanging gases with the water. While submerged, these turtles can gather about 40 percent of their total oxygen requirements from the water column and void about 80 percent of their carbon dioxide waste product into it. This allows these turtles to dive for much longer periods than Alabama's other turtles. After prolonged dives, softshells can repay their oxygen debt with one or two deep breaths rather than the many shallow breaths of other turtles (Stone et al. 1992).

Key to the Species of *Apalone* of Alabama

1a Leading edge of carapace lacking tubercles or other prominences. Go to **2**.

1b Leading edge of carapace with tubercles. Go to **3**.

From left to right:

Head, neck, and anterior carapace of Midland Smooth Softshell lacking tubercles

Head, neck, and anterior carapace of Florida Smooth Softshell with tubercles

2a Top of snout lacking stripes; carapace of juveniles and adult males with large circular spots; mature females of this and the following species often virtually indistinguishable.

***Apalone calvata*—Gulf Coast Smooth Softshell.....page 65.**

2b Top of snout usually with stripes; carapace of juveniles and adult males with scattered dark dashes and pale spots.

***Apalone mutica*—Midland Smooth Softshell . . . . page 68.**

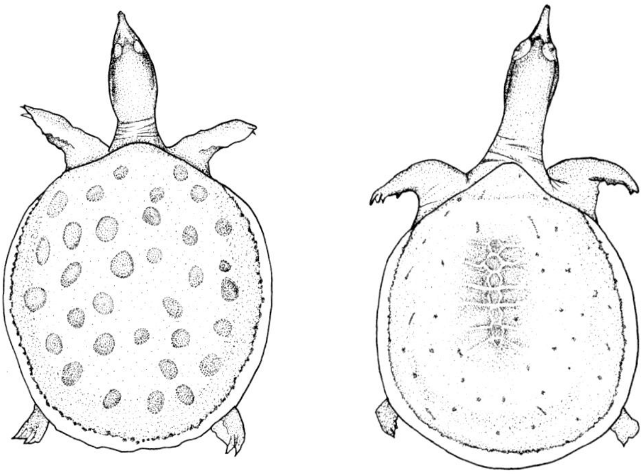

From left to right:

Carapace of Gulf Coast Smooth Softshell with large circular spots

Carapace of Midland Smooth Softshell with dark dashes and pale spots

3a Leading edge of carapace with a marginal ridge; anterior carapacial tubercles low, rounded.

Apalone ferox—**Florida Softshell . . . . page 71.**

3b Leading edge of carapace lacking a marginal ridge; anterior carapacial tubercles often pointed or spine-tipped.

Apalone spinifera ssp.—**Spiny Softshells . . . . page 74.**

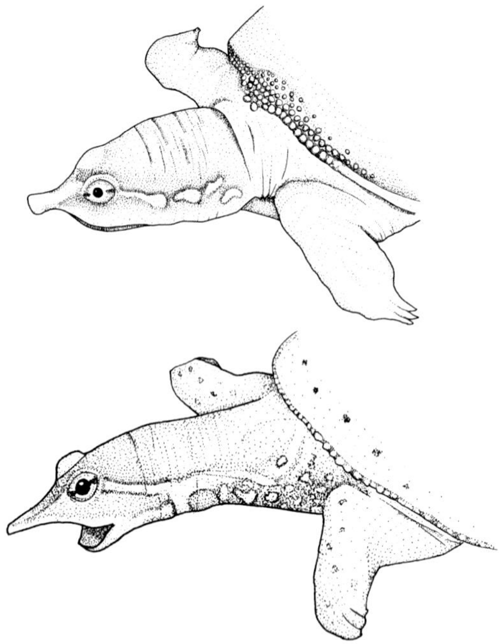

From top to bottom:

Head, neck, and anterior carapace of Florida Smooth Softshell with marginal ridge

Head, neck, and anterior carapace of Spiny Softshell without marginal ridge

Gulf Coast Smooth Softshell

Apalone calvata (Webb, 1959)

Description This species is a moderately large aquatic turtle attaining a maximum carapace length of 11.5 inches (287 mm) in females (males are about half the size of females). The body is flattened and the carapace and plastron are covered with soft skin. The head is elongate because the snout ends in a tubular proboscis. Septal ridges in nostrils are lacking, so that a single opening appears at the tip of the proboscis. The top of the head and snout lack light stripes, a feature that distinguishes this species from *A. mutica*. The leading edge of the carapace lacks tubercles. In adults, the carapace is grayish to light brownish with a light border. In juveniles and males, the dorsum has large circular spots (dusky spots and short lines or dashes in *A. mutica*); these are replaced by obscure mottling or irregular blotches in adult females. The head has a bold, dark-bordered light stripe behind the eyes and inconspicuous light stripes along the shout in front of the eyes. This dark border to the light stripe is wider than in *A. mutica*. These taxonomic characters will not distinguish large females of *A. calvata* from females of *A. mutica*.

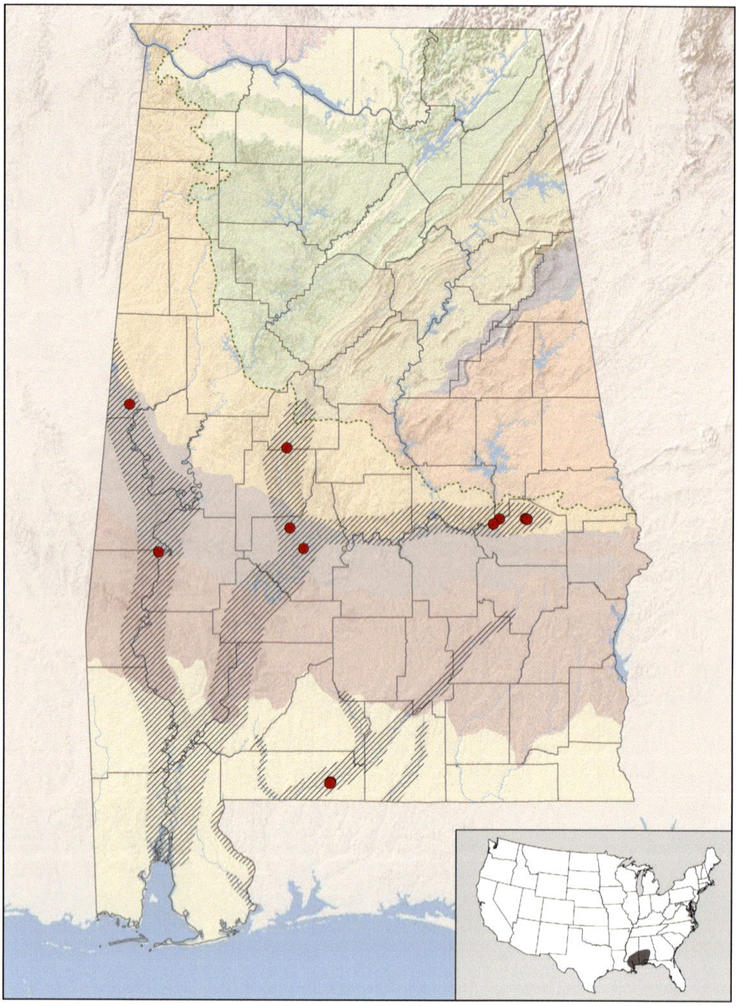

Right: Alabama and continental US distribution of the Gulf Coast Smooth Softshell

Previous: Juvenile Gulf Smooth Soft-shell, Escambia County, FL

Alabama Distribution In Alabama this species is known only from the Coastal Plain in the Alabama, Cahaba, Conecuh, and Tallapoosa Rivers. A Mississippi record from the Tombigbee River establishes the presence of *A. calvata* in that river system.

Habits Gulf Coast Smooth Softshells are confirmed stream dwellers and are strongly aquatic, forsaking the water only to nest or to bask along the edge. Rapid swimmers, they would be difficult to collect were it not for the habit of burying themselves in sand in shallow water. A slightly mounded disturbed area often reveals a turtle's presence. As with other *Apalone, A. calvata* is almost wholly carnivorous, feeding on such items as crayfish, insects, mollusks, fish, frogs, and tadpoles. In contrast to *A. ferox* and *A. spinifera, A. calvata* is not aggressive when handled. Nesting takes place during May, June, and July. The flask-shaped nest cavity is usually dug on an exposed sandy bank or sand bar. The eggs, which reportedly number from four to thirty-three per clutch, are spherical and have hard shells. In Louisiana, nests of this species tend to be positioned on east-facing sand bars and are placed on steeper slopes and contain fewer eggs (average 6.7 eggs/clutch) than *A. spinifera aspera* (Doody 1995).

CONSERVATION AND MANAGEMENT Like all North American Softshells, the Gulf Coast Smooth Softshell is harvested for meat, which is used in soups and stews by local Alabamians or is exported to Asia for sale. Additionally, trade in turtles as pets creates a second market that is particularly strong in Europe. Perhaps because of hunting pressures associated with these markets, this species is not abundant. For these reasons, commercial harvesting is banned. Only two turtles per day may be possessed for personal use. Restriction of activities, such as use of off-road vehicles, on sand bars during nesting season is likely to improve populations; dredging of sandbars and modification of stream channels with riprap are likely to decrease populations of this species.

TAXONOMY Typically, this species is considered to be a subspecies of *A. mutica*. However, we elevate it to species status because of its distinctive color pattern and geographic isolation. Data from the mitochondrial genome examined by Weisrock and Janzen (2000) suggest that *A. calvata* extends from the rivers entering Lake Maurepas in Louisiana, flowing east to include the rivers entering Mobile and Pensacola Bays. Its closest relative, the Midland Smooth Softshell, *A. mutica*, inhabits the Mississippi River and its tributaries. The two overlap only in an extremely narrow zone in southeastern Louisiana. By any species concept, these taxa ought to be recognized as separate species. Previous authors have placed this species in the genera *Amyda, Apidonectes, Potamochelys,* and *Trionyx*.

Midland Smooth Softshell

Apalone mutica (Le Sueur, 1827)

Description This species is a moderately large aquatic turtle attaining a maximum carapace length of about 14 inches (355 mm) in females and about 7 inches (175 mm) in males. The body is flattened and the carapace and plastron are covered with soft skin. The head is elongate because the snout ends in a tubular proboscis. Septal ridges in the nostrils are lacking, so that a single opening appears at the tip of the proboscis. The top of the snout has light stripes, a feature lacking in large females and in all members of its most similar species, *A. calvata*. The leading edge of the carapace lacks tubercles. In adults, the carapace is grayish to light brownish with a light border. In juveniles and males, there is a pattern of dusky spots and short lines or dashes, becoming obscure or absent in some adult males and being replaced by a pattern of mottling or irregular blotches in adult females. The head has a bold, dark-bordered light stripe behind the eyes and has inconspicuous light stripes along the snout in front of the eyes.

Alabama Distribution This species is confined to the Tennessee River drainage, where it occurs in the main channel of the river, its backwaters, and possibly in some of its larger tributaries.

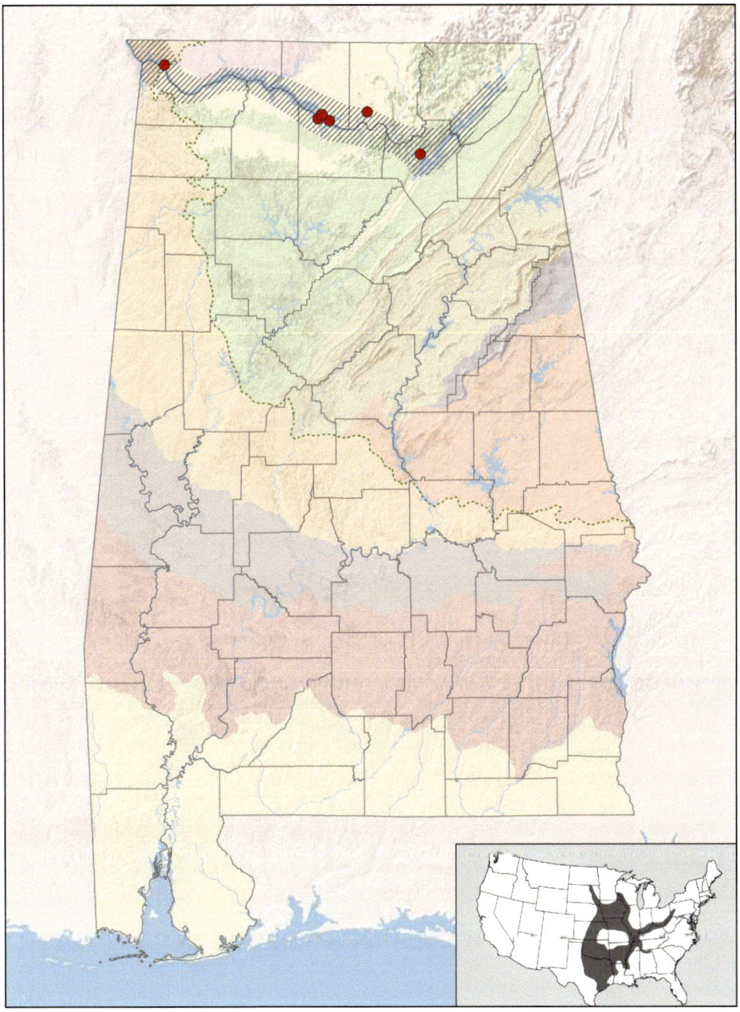

Left: Alabama and continental US distribution of the Midland Smooth Softshell

Opposite: Adult Midland Smooth Softshell, Dane County, WI

Habits The Midland Smooth Softshell is a confirmed stream dweller and is strongly aquatic, forsaking the water only to nest or to bask along the edge. Rapid swimmers, they would be difficult to collect were it not for the habit of burying themselves in mud in shallow water. A slightly mounded, disturbed area often reveals a turtle's presence. As with other *Apalone*, *A. mutica* is almost wholly carnivorous, feeding on such items as crayfish, insects, mollusks, fish, frogs, and tadpoles. In contrast to *A. ferox* and *A. spinifera*, *A. mutica* is not particularly aggressive when handled. Nesting takes place during May, June, and July. The flask-shaped nest cavity is usually dug on an exposed sandy bank or sand bar. The eggs, which reportedly number from four to thirty-three per clutch, are spherical and have hard shells.

Conservation and Management Like all North American Softshells, the Midland Smooth Softshell is harvested for meat, which is used in soups and stews by local Alabamians or is exported to Asia for sale. Additionally, trade in turtles as pets creates a second market that is particularly strong in Europe. Perhaps because of hunting pressures associated with these markets, Midland Smooth Softshells are not abundant. For these reasons, commercial harvesting of this species is banned. Only two turtles per day may be possessed for personal use. Restriction of activities on sand bars during the nesting season is likely to improve populations; dredging of sandbars and modification of stream channels with riprap are likely to reduce habitat for these turtles.

Taxonomy This species typically is conceived as containing two subspecies. However, the two differ in color pattern and have an extremely limited zone of overlap. By any species concept these taxa ought to be recognized as separate species, and we take this action. Data from the mitochondrial genome examined by Weisrock and Janzen (2000) suggest that *A. mutica* inhabits the Mississippi River and its tributaries. It is the sister taxon of *A. calvata*, the Gulf Coast Smooth Softshell, which is distributed from the rivers entering Lake Maurepas in Louisiana, flowing east to include the rivers entering Mobile and Pensacola Bays. Previous authors have placed *A. mutica* in the genera *Amyda, Apidonectes, Potamochelys,* and *Trionyx.*

Florida Softshell

Apalone ferox (Schneider, 1783)

Description This species is a large aquatic turtle attaining a maximum carapace length of about 18 inches (460 mm) in females and 11 inches (285 mm) in males. The body is flattened, slightly longer than wide and slightly narrower at the anterior than the posterior end. Like all North American Softshells, the head is elongate, ending in a snout that is prolonged into a tubular proboscis. The shell is covered with soft, leathery skin, and the leading edge has a series of low rounded protuberances. Along the anterolateral edge of the carapace the skin folds continuously, creating an appearance similar to tuck-and-roll upholstery. The carapace of juveniles has a light margin with a series of conspicuous dark blotches, sometimes with light centers. The carapace pattern becomes obscure with age. Large females have enlarged, flattened knobs in the nuchal region of the carapace.

Alabama Distribution Florida Softshells are restricted to the Lower Coastal Plain, where they occur locally from the extreme southeastern corner of the state, westward along the southern tier of counties, to Mobile Bay.

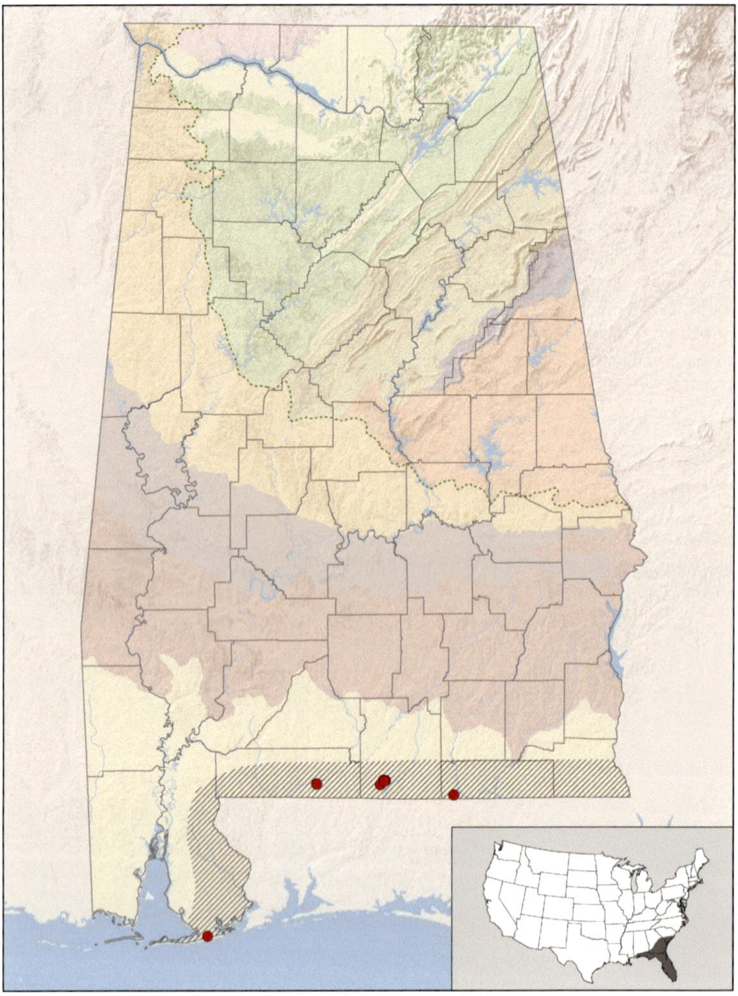

Right: Alabama and continental US distribution of the Florida Softshell

Previous: Adult Florida Softshell, Covington County, AL

Habits This species, an inhabitant of sluggish streams, lakes, and ponds, spends much of its time buried on the substrate with only the head protruding. It feeds on a wide variety of animal life, including crayfish, insects, mollusks, frogs, and fish. As is the case with other softshells, this species is rather thoroughly aquatic and leaves the water only to bask, a rare event, or to lay eggs. Out of the water it is extremely pugnacious and, because of its long neck, difficult to handle without being bitten. Two or possibly three clutches of seven to twenty-two eggs each are laid in a flask-shaped cavity constructed by a female in sandy soil.

Conservation and Management Florida Softshells are consumed by some locals and are harvested for export to Asia. Additionally, this species is sold in the pet trade, largely to European countries. These two markets put such pressure on native populations that Alabama populations are protected as a Priority 3 (Moderate Conservation Concern) species. Therefore, it is unlawful to possess this species without a special permit. Because ponds are relatively common in south Alabama, habitat for these turtles is abundant. But nesting areas should be managed to reduce disturbances caused by off-road vehicles during the nesting season. Additionally, predator control (especially raccoons and feral hogs) may be necessary to maintain stable populations. Finally, because members of this species, especially the females, cross roads frequently (Steen et al. 2006), wildlife fencing and designed underpasses in areas of heavy traffic may be necessary to maintain this species in urban settings.

Taxonomy No subspecies are recognized for this species, and no distinctive mitochondrial lineages were discovered from samples across its geographic range (Weisrock and Janzen 2000). This species is the closest living relative to *A. spinifera* (Weisrock and Janzen 2000). Previous authors have placed this species in the genera *Amyda, Aspidonectes, Platypeltis, Testudo,* and *Trionyx.*

Spiny Softshells

Apalone spinifera (Lesueur, 1827)

Taxonomy This species has the broadest geographic range of all North American softshells. It is most closely related to *A. ferox* (Weisrock and Janzen 2000). Seven subspecies are recognized, two of which occur in Alabama. McGaugh et al. (2008) document that the four western subspecies, *A. s. atra, A. s. emoryi, A. s. guadalupensis,* and *A. s. pallida,* are deeply divergent from the three eastern subspecies, *A. s. aspera, A. s. hartwegi,* and *A. s. spinifera.* However, they stopped short of recognizing these as two distinct species. For the eastern subspecies, these authors found no support for retaining *A. s. hartwegi* as being distinct from *A. s. spinifera* but found a deep division between *A. s. spinifera* and *A. s. aspera,* the two subspecies found in Alabama. Again, these authors stopped short of recognizing these as distinct species in part because one specimen of *A. s. aspera* from the Suwannee River in Georgia was documented to be basal to the otherwise monophyletic sister taxa *A. s. aspera* and *A. s. spinifera.* Because the Georgia specimen was morphologically referable to *A. s. aspera* (incorrectly listed as *A. s. emoryi* in McGaugh et al. 2008), we retain *A. s. aspera* and *A. s. spinifera* as subspecies but restrict *A. s. aspera* to specimens from the Gulf Coast and assume the Suwannee specimen to represent a currently unrecognized taxon. If further collections confirm this pattern, we recommend elevating all eastern subspecies to specific status. Alabama specimens from Uphappee Creek of Macon County and the Conecuh River of Covington County document the genetic lineage *A. s. aspera* from the state, but no Alabama specimens of *A. s. spinifera* were included in the study of McGaugh et al. (2008). Previous authors have placed this species in the genera *Amyda, Aspidonectes, Gymnopus, Platypeltis, Trionyx,* and *Tyrse.*

Key to the Subspecies of *Apalone spinifera* of Alabama

1a Rear of carapace with one encircling dark line.

Apalone spinifera spinifera—**Eastern Spiny Softshell . . . . page 76.**

1b Rear of carapace with two or more encircling dark lines.

Apalone spinifera aspera—**Gulf Coast Spiny Softshell . . . . page 79.**

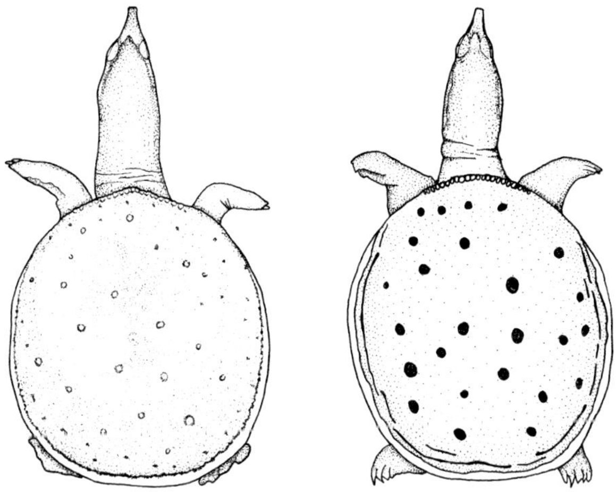

From left to right:

Carapace of Eastern Spiny Softshell with one encircling dark line

Carapace of Gulf Coast Spiny Softshell with multiple encircling dark lines

Eastern Spiny Softshell

Apalone spinifera spinifera (Le Sueur, 1827)

Description These are large aquatic turtles that attain a maximum carapace length of about 17 inches (430 mm) in females and about half that size in males. Their body is flattened and nearly round. The head is elongate, ending in a snout with a tubular proboscis. Right and left nostrils are apparent because of a visible nasal septum. The carapace and plastron are covered with soft, pliable skin. At the leading edge of the carapace are acute tubercles or spiny projections, and its trailing edge has a single encircling dark line inside the margin. In juveniles and males the carapace has well-defined dark ocelli (light center) and spots (dark center). In adult females the carapace is irregularly mottled with dark and light markings, which contrast more sharply than in *A. s. aspera*. The undersurfaces of the legs and edges of plastron are conspicuously mottled. In large females, the carapace has enlarged flattened knobs in the nuchal region and posteriorly in the center of the shell. The carapace of adult males has a rough texture similar to sandpaper.

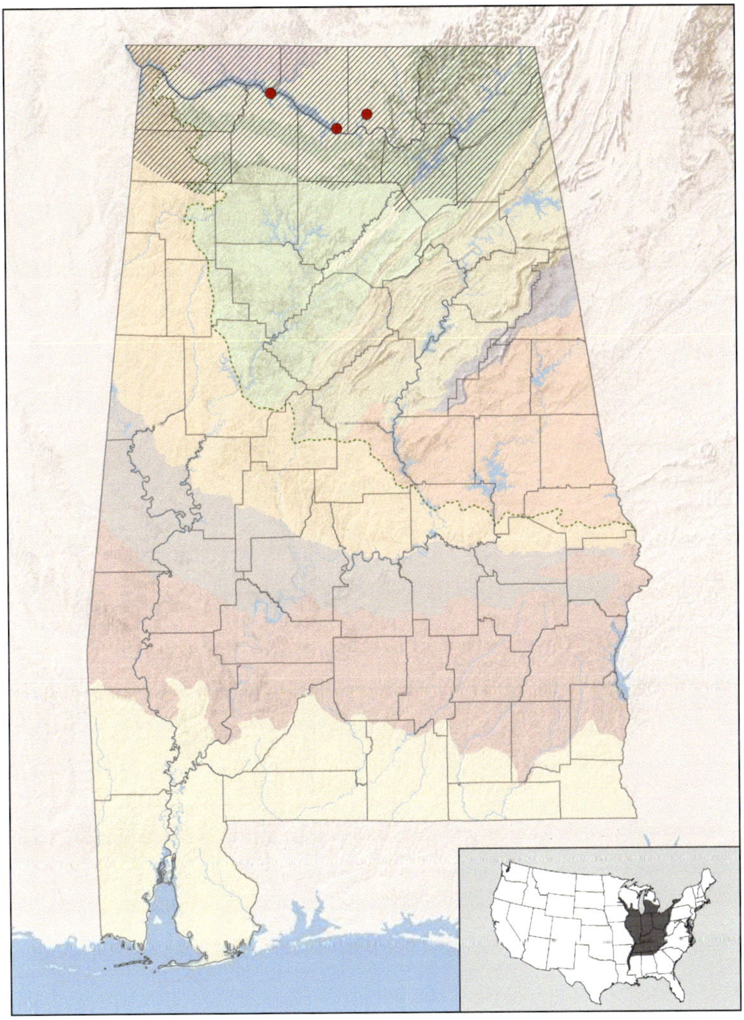

Left: Alabama and continental US distribution of the Eastern Spiny Softshell

Opposite: Juvenile Eastern Spiny Softshell, Madison County, AL

Alabama Distribution This subspecies is found in streams of the Tennessee River system and possibly some ponds and lakes in the area drained by those streams.

Habits The habits of the Eastern Spiny Softshell in Alabama are not well known. The optimal habitat of this form in Alabama is a free-flowing creek or river with a sand or gravel substrate. The Midland Smooth Softshell, *A. mutica,* a turtle with which it is sympatric in Alabama, seems best adapted to streams or backwaters of streams having deposits of silt, or sand and silt, on the bottom and along the margins. These turtles are pugnacious and willing to bite when handled. Therefore, it is advisable to grasp them with both hands by the rear of the carapace. Food items include a wide variety of invertebrates and small vertebrates. The nesting season begins in May and ends in July, during which three to twenty-five eggs are laid in flask-shaped nests constructed in exposed sand bars and beaches. Females probably lay up to three clutches each season.

Conservation and Management The impoundment of the Tennessee River throughout its length in Alabama has apparently been detrimental to the Eastern Spiny Softshell. Additionally, like all softshells, the Eastern Spiny Softshell is harvested for meat, which is used in soups and stews by local Alabamians or is exported to Asia for sale. Trade in turtles as pets creates a second market that is particularly strong in Europe. Perhaps because of hunting pressures associated with these markets, this subspecies is not abundant. For these and other reasons, commercial harvesting of this subspecies has been banned. Only two turtles per day may be possessed for personal use. Restriction of activities on sand bars during nesting season is likely to improve populations; dredging of sandbars and modification of stream channels with riprap are likely to decrease populations.

Gulf Coast Spiny Softshell

Apalone spinifera aspera (Agassiz, 1857)

Description This subspecies is a large aquatic turtle reaching a maximum carapace length of about 18 inches (455 mm). The body is flattened and nearly round with a carapace and plastron that are covered with soft skin. The plastron is relatively short, compared to that of *A. s. spinifera*. The head is elongate, ending in a snout with a tubular proboscis. Right and left nostrils are apparent because of a visible nasal septum. At the leading edge of the carapace are acute tubercles or spiny projections, and its trailing edge has two or more dark lines encircling the inside of the margin (single dark line in *A. s. spinifera*). In juveniles and males the carapace has well-defined dark ocelli (light center) and spots (dark center). In adult females the carapace is uniform or irregularly mottled with dark and light markings, which contrast weakly with each other (strong contrast in *A. s. spinifera*). In large females, the carapace has enlarged flattened knobs in the nuchal region and posteriorly in the center of the shell. The carapace of adult males has a rough texture similar to sandpaper.

Alabama Distribution This subspecies is widespread in rivers, creeks, lakes, and permanent ponds throughout Alabama except for that portion drained by the Tennessee River and its tributaries, where it is replaced by *A. s. spinifera*.

Above: Adult Gulf Coast Spiny Softshell, Decatur County, GA

Previous: Juvenile Gulf Coast Spiny Softshell, Winston County, AL

Habits The Gulf Coast Spiny Softshell is by far the most common softshell in Alabama. This turtle exploits most kinds of permanently aquatic habitats, ranging from small creeks and farm ponds to the largest rivers and lakes. It is ill tempered when out of water, and a large individual can deliver a painful bite. In handling a large specimen, it is advisable to grasp it with both hands by the rear of the carapace. However, be prepared to be gouged by the toenails of the hind limbs. *Apalone s. aspera* is carnivorous and seems to eat almost anything available in the way of animal life. The dominant food items are invertebrates, but small vertebrates, including fish, are consumed; there is no evidence that this turtle harms fish populations. The nesting season begins in May and ends in July. From three to twenty-five eggs are laid in flask-shaped nests constructed in sandy places. Exposed sand bars and beaches are especially favored as nesting sites. The females probably lay up to three clutches each season.

Conservation and Management Flesh from these animals is used in soups and stews and animals captured in the United States are exported to Asia for sale. Additionally, trade in turtles as pets creates a secondary market that is particularly strong in Europe. Because of

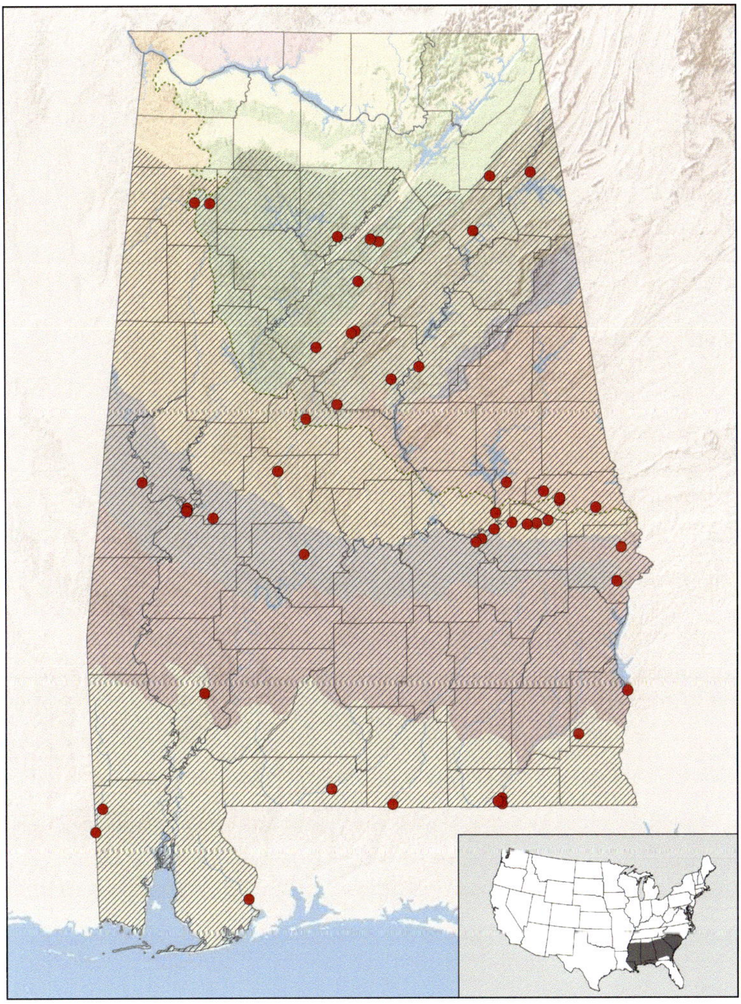

Alabama and continental US distribution of the Gulf Coast Spiny Softshell

these pressures, commercial harvesting of this subspecies is banned. Only two turtles per day may be possessed for personal use. Restriction of activities on sand bars during the nesting season is likely to improve populations; dredging of sandbars and modification of stream channels and shorelines with riprap are likely to decrease populations of Gulf Coast Spiny Softshells.

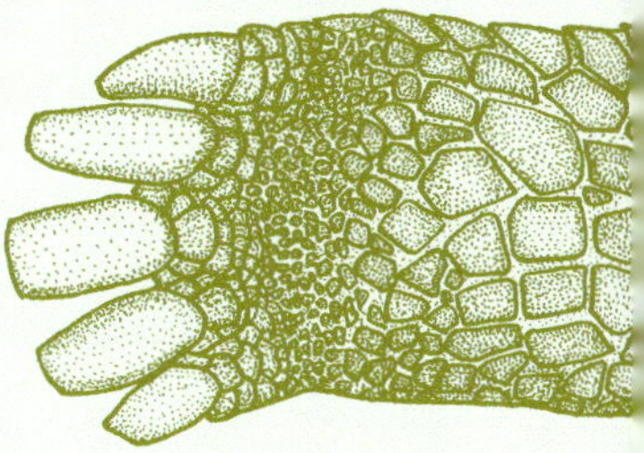

Tortoises

Family Testudinidae

Tortoises occur in North and South America, Africa, Europe, and Asia. Fifteen genera and approximately forty-eight species are recognized. These turtles have appendages that lack webbing between the digits and therefore are adapted for terrestrial rather than aquatic locomotion. The family Testudinidae is most closely related to the family Geoemydidae (Shaffer et al. 2008), a group of aquatic turtles from the New and Old Worlds. Therefore, tortoises represent a radiation of turtles that evolved characteristics which allowed them to leave an aquatic environment and invade dry, savanna habitats. This divergence event is estimated to have occurred 52 million years ago (Near et al. 2005). The group is noted for achieving great age in nature, even within a turtle radiation that is known for great longevity relative to similarly sized tetrapods (Gibbons 1987). One genus, *Gopherus,* occurs in the United States, and this is the one that occurs in Alabama. However, the fossil record documents the genus *Cymotholcus* to have been present in Alabama about 35 million years ago and the genera *Floridemys, Geochelone,* and *Hesperotestudo* to have been present in the southeastern United States. These were all much larger than *Gopherus* and were contemporaneous with this small genus until about 100,000 years ago.

North American Tortoises

Genus *Gopherus* (Rafinesque, 1832)

This genus contains four species of relatively small tortoises found in the United States and northern Mexico. All species in the genus live in open areas of savanna or deserts where they survive with limited access to water. The elephantine hind feet and forelimbs designed for digging make this genus immediately identifiable compared to all other turtles in Alabama. The genus is most closely related to the Asian genus *Manouria,* suggesting that the ancestor to all North American tortoises crossed the Bering land bridge to reach the continent (Le et al. 2006). Only one species currently occurs in Alabama.

Gopher Tortoise

Gopherus polyphemus (Daudin, 1802)

Description Gopher Tortoises are large terrestrial turtles attaining a maximum carapace length of about 14 inches (355 mm). The top of the head is conspicuously scaled compared with all other turtles in the state, and the front toenails are large, flattened, and designed for digging. The hind feet lack webbing between the toes and instead have round pads that form the sole of the foot, making it like an elephant's. The gulars of the plastron protrude well forward, creating a slightly forked structure below the neck. The carapace is brown in adults but contains scutes that are yellow in the center in young. The skin is yellowish-brown in juveniles, darkening to brown or almost black in adults. The plastron of adult males has a distinct, centralized concavity, while that of females is flat or has a wide concavity.

Alabama Distribution This species is restricted to deep sandy soils of the Coastal Plain. It can be common and relatively uniformly distributed in the Lower Coastal Plain, becoming local with low density toward the northern extent of its range. It occurs in the Fall Line Hills, as evidenced by a small remnant population in Russell County. It is

Above: Adult Gopher Tortoise and burrow, Wayne County, MS

Previous: Hatchling Gopher Tortoise, Covington County, AL

absent from the Black Belt but is present in the Red Hills, occupying pockets of longleaf pine forest on the tops of ridges. At least one population, a small one occupying portions of Pike and Bullock Counties, occurs in the transitional zone between the Black Belt and Red Hills. Harper (1943) describes apparently abandoned burrows from "a few miles east of Selma, on the north side of the River," suggesting that gopher tortoises might have been present farther north, before succumbing to conversion of land to agriculture and predation by humans.

Habits Gopher Tortoises live in dry, sandy places, where they construct sloping burrows up to 30 feet (9 m) in length and 15 feet (4.5 m) deep.

Habitats in Alabama capable of supporting Gopher Tortoise populations are usually dominated by turkey oak and longleaf pine. Gopher Tortoise burrows provide shelter not only for the Gopher Tortoises, but for over 350 other animals as well (Jackson and Milstrey 1989), including rare species such as Gopher Frogs, Florida Pinesnakes, Eastern Indigo Snakes, and Eastern Diamond-backed Rattlesnakes. Additionally, tortoises disrupt understory vegetation when they dig their burrows, an activity that plays a role in maintaining an unusually diverse assemblage of flowering plants (Kaczor and Harnett 1990).

Above: Burrow of a Gopher Tortoise, Covington County, AL

The Gopher Tortoise is herbivorous, feeding on a variety of herbs and fruits. A staple food at many localities is wiregrass (*Aristida beyrichiana* and *Sporobolus junceus*), but a wide variety of other grasses and forbs are consumed. In fact, foraging activities of gophers likely alter the plant assemblages associated with gopher burrows and the trails that tortoises use to move from one burrow to another (Boglioli et al. 2000; Birkhead et al. 2005).

Courtship in the Gopher Tortoise involves nipping and head-bobbing (Auffenberg 1966), occurring throughout the season of activity and peaking in August and September, near the end of the activity season (Johnson et al. 2008a) when males scramble to find receptive females (Johnson et al. 2009). Nesting occurs from May to July and may involve fertilization from sperm stored from mating with several males the previous season (Moon et al. 2006). Between four and seven round, hard-shelled eggs are laid in a cavity approximately 6 inches (15 cm) beneath the surface (Carr 1952; Arata 1958). The nest hole is usually dug within a few feet of a burrow entrance and the female may abandon that burrow for the rest of the activity season (Ott et al. 2003). The sex of offspring from these eggs is determined by the temperature of the nest, with females produced from warmer nests and males from colder ones (Burke et al. 1996).

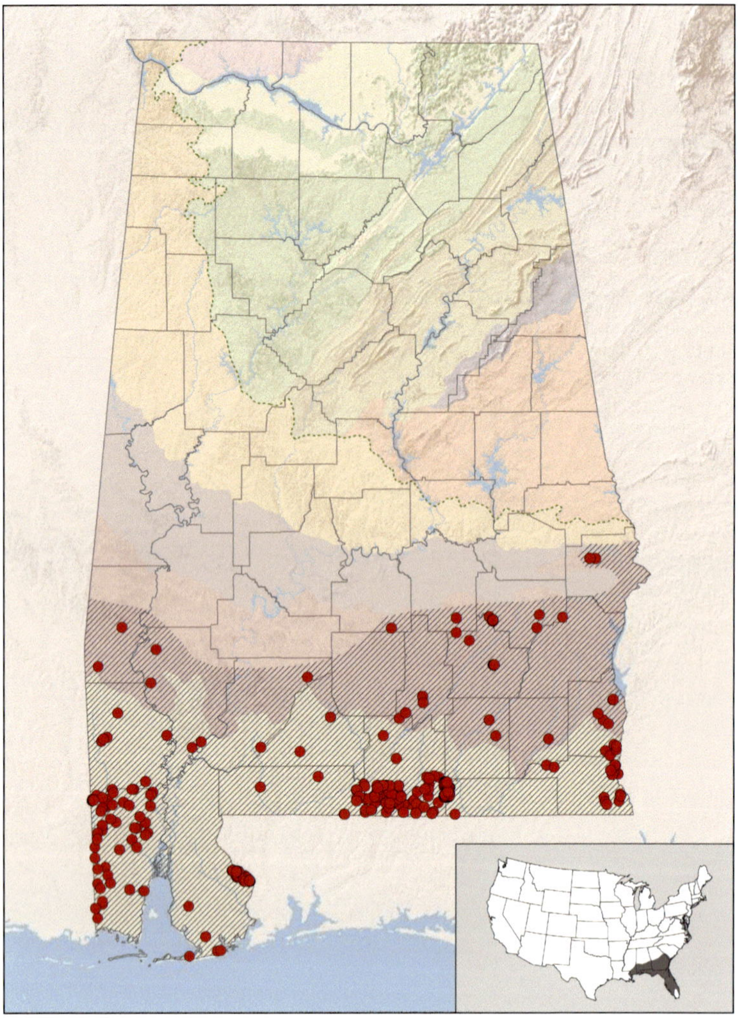

Alabama and continental US distribution of the Gopher Tortoise

Gopher Tortoises live in large aggregations occupying ridges created by deep, excessively well-drained, sandy soils; here, individuals select the most open areas possible to construct their burrows (Guyer and Hermann 1997). Such aggregations are referred to as colonies, and as long as the habitat is burned frequently and maintained in an open, park-like state, tortoises will maintain a set of burrows for decades (Guyer and Hermann 1997). In this setting, neighbors likely know each other as individuals and visit each other on a daily basis in the active season. This social setting requires remarkable spatial memory of burrows, the landscape along trails leading to burrows, and the size and shape of individuals. Recent analyses of social networks within tortoise colonies document that tortoises form cliques of individuals that have strong behavioral ties to each other but not to members of other cliques. These cliques overlap in space, so the maintenance of this social structure requires avoidance behaviors. When placed in the context of social organization of other vertebrates, this may represent the most complex social organization of any turtle (Guyer et al. 2012).

Conservation and Management Gopher Tortoises west of the Tombigbee River were listed as threatened by the USFWS in 1987 and therefore are federally protected under the ESA. In the remainder of its distribution in Alabama it is ranked Priority 2 (High Conservation Concern) by ADCNR (Mirarchi et al. 2004) and, therefore, is protected by the state of Alabama. Strong evidence suggests that this species is declining even on public lands set aside for wildlife and managed to enhance the habitat for declining species. This suggests that current habitat management practices are insufficient to maintain optimal conditions (McCoy et al. 2006). In Alabama, conservation problems have been exacerbated by persistent gassing of Gopher Tortoise burrows by humans attempting to collect snakes for the annual Opp Rattlesnake Rodeo. In 2009, such gassing of burrows was made illegal, but the environmental legacy of this practice likely remains across southern Alabama. For these reasons the Gopher Tortoise is being considered for protection under the ESA throughout the entire range of the species. Such action appears warranted for Alabama populations because, of the two million tortoises that likely once occupied the state, only about 45,000 to 175,000 remain (Guyer et al. 2011).

A recovery plan for the species was filed in 1990 for populations receiving federal protection. This plan focuses on preventing further endangerment and achievement of a benchmark for population recovery that would allow delisting. The DeSoto National Forest in southeastern Mississippi is the focus of plans for preventing further endangerment, and a target of 22,400 individuals on 7,343 ha of suitable habitat is given as the recovery goal. Recent work on population genetics has documented the limited genetic diversity in the DeSoto National Forest (Ennen et al. 2010), likely because of a genetic bottleneck in the recent past. It is also likely that this bottleneck extended to portions of extreme western Alabama. Alabama's role in the recovery plan is limited to participating in achieving a density of approximately one tortoise burrow per acre (three per ha) in the 45,927 acres (18,594 ha) of private lands with deep sandy soils within the listed range, a second target set for achieving delisting.

One conservation strategy affecting Gopher Tortoises in Alabama is the practice of moving individuals from lands scheduled for development to lands managed as tortoise reserves. For this plan to work, the reserves must be of sufficient size to retain a viable population of Gopher Tortoises. Current data suggest that a viable population consists of 200–400 individuals that, in current heavily managed landscapes, are distributed across 3,460 acres (1400 ha; McCoy and Mushinsky 2007; Styrsky et al. 2010). Because tortoise density increases with increasing habitat quality, reserve areas could be much smaller than this estimate if habitat management efforts are aggressive enough to create and maintain the open habitat that Gopher Tortoises prefer. We know of three translocated breeding populations, one each in Autauga, Bullock, and Macon Counties. Three additional reserves have been created in Mobile County by state agencies, one by the Alabama Department of Transportation, one by South Alabama Utilities, and one by the Mobile Area Water and Sewage System. The areas of these reserves range from 220 to 600 acres (90–243 ha), which means they will need to be managed exceptionally aggressively to perform their function as conservation reserves. In addition to these reserves, the State Lands Division of ADCNR has created a population of Gopher Tortoises at the Wehle Tract in Bullock County, and the state wildlife action plan recommends using power line rights of way as potential conservation space for Gopher Tortoises (Alabama Department of Conservation and Natural Resources, Division of Wildlife and Freshwater Fisheries 2005).

Habitat management goals to enhance forests for Gopher Tortoises are fairly straightforward. Standing timber needs to be thinned while leaving enough adult pine trees to produce pine straw of sufficient abundance to carry frequent (every other year) fire. Growing-season fires are preferred because these enhance desirable vegetative growth, especially of warm-season grasses that also provide fuel for fire. These are the conditions that allow gopher tortoises to maintain the intricate social ties that characterize the best remaining colonies (Guyer et al. 2012).

Taxonomy No subspecific variation is recognized for this species, but three distinctive lineages have been identified by analysis of mtDNA segments (Osentoski and Lamb 1995). An eastern and a western lineage abut along the Apalachicola River drainage, and a central Florida lineage is associated with the Brooksville Ridge. All Alabama specimens appear to belong to the western lineage. Additional unpublished studies of genetic variation demonstrate that major rivers likely serve as dispersal barriers causing genetic divergence among Gopher Tortoise populations. Therefore, Alabama likely has two major genetic lineages, one on each side of Mobile Bay. Previous authors have placed this species in the genus *Testudo*.

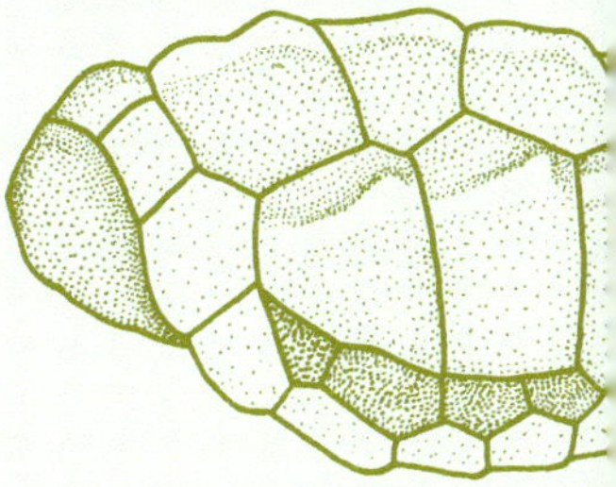

Snapping Turtles

Family Chelydridae

This family of large freshwater turtles is found from the eastern United States and the adjacent portion of southern Canada southward to Ecuador. It consists of two genera, both of which are represented in Alabama. Members of this family are immediately recognizable from their large size, rugose shell, and exceptionally long tail. Based on traditional morphological features, Snapping Turtles are the sister taxon to the common ancestor of all other cryptodiran turtles (E. S. Gaffney, pers. comm.). However, emerging molecular data place it as sister to the families Cheloniidae + Dermochelyidae + Kinosternidae (Shaffer et al. 2008). Regardless of which phylogenetic reconstruction is correct, the lineage leading to members of this family is ancient, diverging 85 million years ago according to one implementation of a molecular clock (Near et al. 2005).

Key to the Genera of Chelydridae of Alabama

1a Carapace without a row of supramarginal scutes; eyes directed dorsolaterally; snout moderately hooked.

Genus ***Chelydra*****—Eastern Snapping Turtle . . . . page 96.**

1b Carapace with a row of supramarginal scutes between the marginals and pleurals; eyes directed laterally; snout strongly hooked.

Genus ***Macrochelys*****—Alligator Snapping Turtle . . . . page 100.**

From top to bottom:

Carapace of Eastern Snapping Turtle without row of supramarginal scutes

Carapace of Alligator Snapping Turtle with row of supramarginal scutes

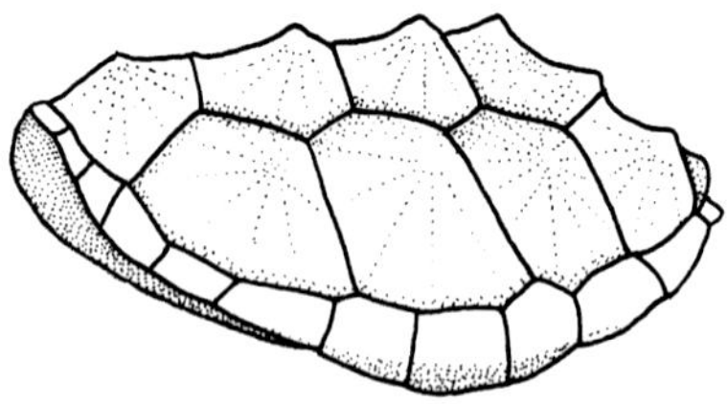

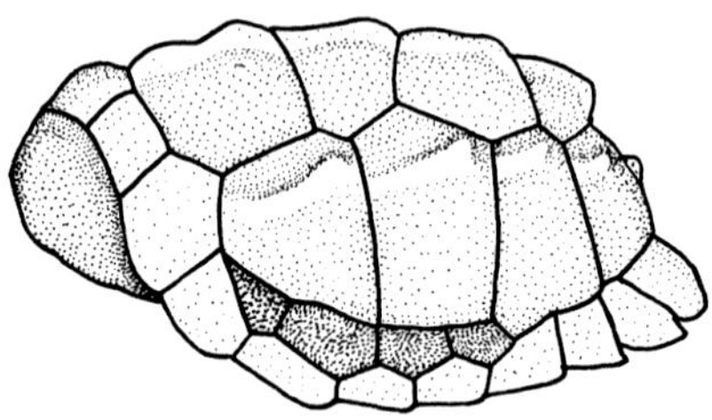

Snapping Turtles

Genus *Chelydra* (Schweigger, 1812)

This genus is wide-ranging and has a geographic distribution that is the same as that of the family. We follow Shaffer et al. (2008) in considering the genus to consist of three species, one of which, *C. serpentina*, is widespread in the United States, including Alabama. This species appears to have expanded its range relatively recently and shows no consistent subspecific variation (Shaffer et al. 2008). Species within the genus are large and highly aquatic, venturing on land only to nest and to seek new aquatic sites. The genus is sister to the Alligator Snapping Turtle genus *Macrochelys* (Shaffer et al. 2008). These two genera are similar in appearance but *Chelydra* lacks supramarginal scutes (present in *Macrochelys*), has low keels on the carapace (tall in *Macrochelys*), possesses a weakly hooked snout (strongly hooked in *Macrochelys*), and has dorsolaterally directed eyes (lateral in *Macrochelys*).

Eastern Snapping Turtle

Chelydra serpentina (Linnaeus, 1758)

Description Eastern Snapping Turtles are large in size, attaining a carapace length of up to 19 inches (494 mm). The carapace of this species has three keels, but these are not as well developed as those of *Macrochelys*, and they often are lacking in old individuals. The eyes of Eastern Snapping Turtles are easy to see when viewed from above because they are rotated dorsally. In front of the eyes, the snout is elongate with a moderately hooked snout. The plastron is an extremely reduced, cross-shaped structure, exposing much skin on the venter. Three barbels are present on the chin and the skin of the rest of the body is warty. The tail is long, being equal in length to that of the carapace, and has three dorsal rows of tubercles, the median of which is conspicuous and platelike.

Alabama Distribution This species is found throughout the state.

Habits In Alabama the Eastern Snapping Turtle inhabits a wide variety of permanently aquatic habitats, ranging from small farm ponds and creeks to the largest lakes and rivers. Only in some of the sandy-bottomed, clear creeks of the Lower Coastal Plain does it appear to

Left: Hatchling Eastern Snapping Turtles, Madison County, AL

Opposite: Adult Eastern Snapping Turtle, Macon County, AL

be scarce or absent. This turtle is highly aquatic but does leave water to lay eggs and to seek new aquatic sites (Steen et al. 2010). When in water it is inoffensive and not disposed to bite; when removed from the water, it is vicious and aggressive, lunging and striking repeatedly at the offender. The long tail makes a convenient handle for carrying, but large specimens may be injured when carried in this manner, and care must be taken to avoid being bitten on the leg.

The eggs are round with soft shells and are laid in a flask-shaped cavity dug by the female, often well away from water. The usual number per clutch is between twenty and thirty. The sex of offspring from these eggs is determined by the temperature of the nest, with females produced from warmer nests and males from cooler nests (Janzen and Paukstis 1991). Eastern Snapping Turtles are omnivorous. Fish, carrion, and vegetable matter constitute the bulk of the diet.

Conservation and Management Eastern Snapping Turtles remain a common and widespread species within Alabama. They inhabit managed ponds and frequent creeks in urban settings. Nevertheless, two sources of adult mortality may eventually reduce populations of snappers in Alabama. First, they have flesh that is prized as a meat for stews and soups. These animals are thus frequently captured for consumption by humans, and adult turtles are the primary focus of

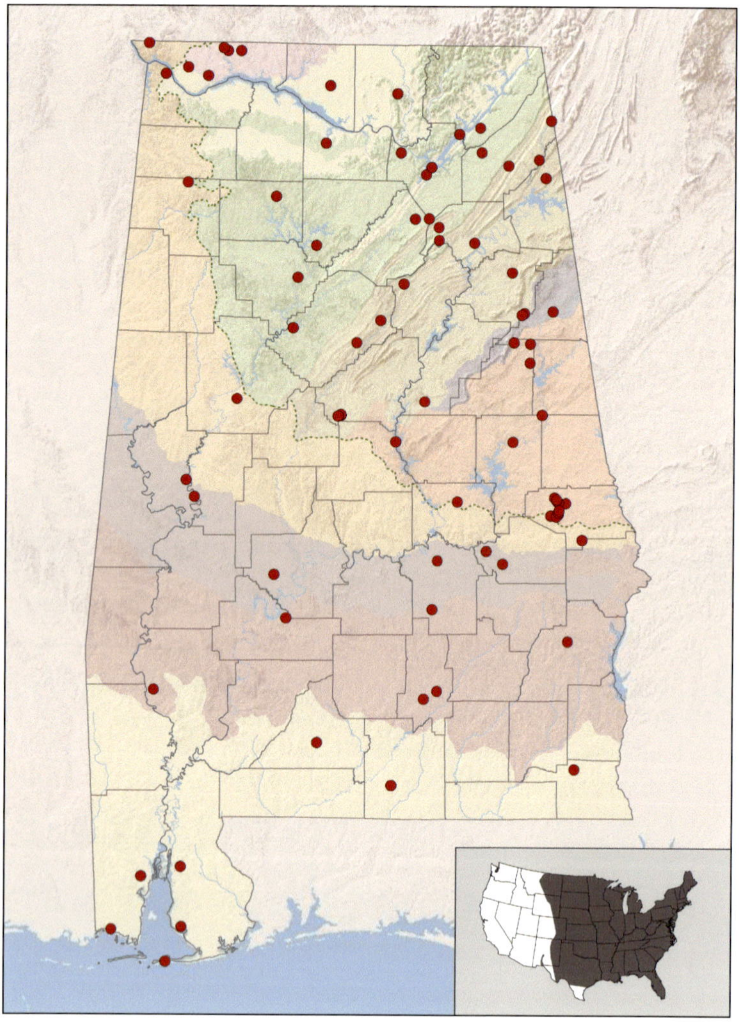

Alabama and continental US distribution of the Eastern Snapping Turtle

this fishery. Second, Eastern Snapping Turtles in areas of high road density suffer increased mortality, especially of females searching for nest sites (Steen and Gibbs 2004).

Taxonomy We follow Shaffer et al. (2008) in recognizing no subspecific variation within this species. Previous publications have placed this species in the genera *Chelonura, Divisia, Emysaura, Macrochelys, Macroclemys,* and *Testudo.*

Alligator Snapping Turtles

Genus *Macrochelys* (Gray, 1855)

This genus contains the Alligator Snapping Turtles of North America, a lineage of extremely large freshwater turtles of the Eastern United States. The genus as typically conceived contains one extant species, which occurs in streams of Gulf drainage systems from northern Florida to Texas, northward to Illinois and Iowa. However, molecular data document a second species restricted to the Suwannee River of Florida and Georgia. This genus is sister to the Eastern Snapping Turtle genus *Chelydra* (Shaffer et al. 2008). These two genera are similar in appearance but *Macrochelys* possesses supramarginal scutes (absent in *Chelydra*), has tall keels on the carapace (short in *Chelydra*), and has laterally directed eyes (dorsolateral in *Chelydra*).

Alligator Snapping Turtle

Macrochelys temminckii (Troost in Harlan, 1835)

Description An extremely large aquatic turtle, this species attains a maximum carapace length of over 25.5 inches (650 mm) for males and about 19.5 inches (495 mm) for females in Alabama (Folt and Godwin 2013). The carapace is brown, rough, and contains three conspicuous longitudinal ridges that may be worn down in old individuals. The carapace generally has a row of two to three supramarginal scutes sandwiched between the marginals and pleurals at the anterior end of each side. The notch located between the posteriormost marginals of the carapace is narrow and triangular or U-shaped. The snout in front of the eyes is elongate and terminates in a strongly hooked upper jaw. When viewed from above, the eyes are directed toward the side of the head and, in contrast with those of *Chelydra,* cannot be seen because of a bony plate that covers the top of the eye socket. The tail is exceptionally long (longer than the length of the carapace). A small, cross-shaped plastron covers only a small portion of the venter with the rest being covered by skin. The skin of the head and neck are rough in appearance because of numerous dermal projections.

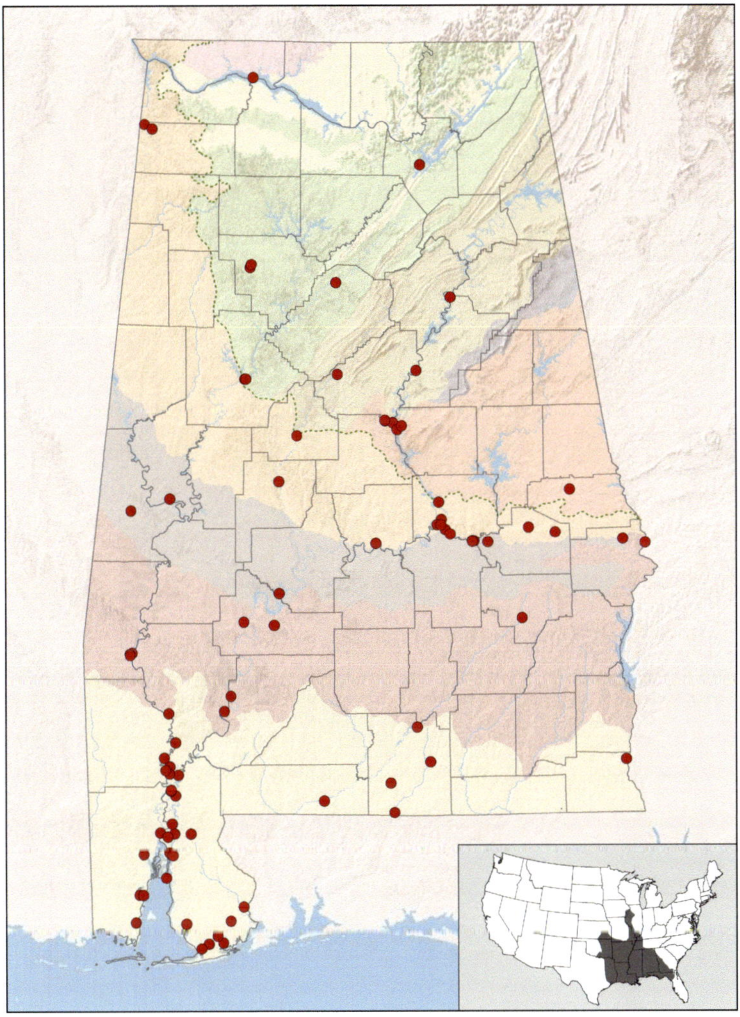

Left: Alabama and continental US distribution of the Alligator Snapping Turtle

Opposite: Juvenile Alligator Snapping Turtle, Elmore County, AL

Alabama Distribution Alligator Snapping Turtles occur throughout the state, with an eastern mitochondrial lineage being restricted to the Chattahoochee, Choctawhatchee, and Pea Rivers and a western mitochondrial lineage being found in all other drainages west of the Pea River.

Habits This spectacular animal, capable of attaining weights in excess of 200 pounds (91 kg), is the world's largest freshwater turtle.

The Alligator Snapping Turtle usually inhabits rivers, oxbows, and sloughs, but is found occasionally in medium-sized creeks. Unlike the Eastern Snapping Turtle, it is almost never found in isolated ponds and lakes and leaves the water only to nest.

Alligator Snapping Turtles are sluggish and spend nearly all their time lying on the bottom of aquatic habitats. From such positions, individuals may stand on the tips of their toes and slowly extend the neck to its full extent in order to reach the water's surface and breathe. Along the coast these turtles may inhabit brackish water for considerable periods. Jackson and Ross (1971) reported barnacles on the shell of a specimen from near the mouth of the Dog River, Mobile County, Alabama, and a similarly infested specimen, from Baldwin County, is known (Mount 1975).

One unique feature of this turtle is a wormlike luring device attached to the tongue. Lying on the bottom, mouth agape, the snapper wiggles this appendage and feeds on fish that are attracted to it. Fish are by no means its sole food. Virtually anything of animal origin, including carrion, is eaten, and some individuals have been found whose stomachs are filled with palmetto fruits or acorns (Mount 1975). This species nests in sandy to loamy soils generally in elevated banks above a river. Females may travel nearly 110 yards (100 m) overland to reach a suitable nest site, and these sites frequently are overgrown with vegetation. The eggs of *M. temminckii* are round with leathery shells that are chalky white and have a rough, sandpaper-like texture. Clutch size ranges from 10 to 50 eggs, the larger females producing larger numbers of eggs. The sex of the offspring from these eggs is determined by the temperature of the nest, with females produced from warmer nests and males from cooler nests (Janzen and Paukstis 1991). Lacking the agility and viciousness of the Eastern Snapping Turtle, the Alligator Snapping Turtle, when removed from water, gapes at its captor. It will bite objects placed in or near the mouth, but the widely accepted idea that it can bite through a broom handle is a gross exaggeration.

Conservation and Management The Alligator Snapping Turtle is ranked Priority 3 (Moderate Conservation Concern) by ADCNR (2012 modification of Mirarchi et al. 2004) and is being considered for protection under the ESA because adult populations were reduced by hunting for the soup industry in the 1940s and 1950s (Pritchard 1989). Although still widely distributed within the state, population densities are lower than in surrounding states, except for a notably dense population in the Fowl River of Mobile County (Folt and Godwin 2013). The species is also protected internationally through its listing as vulnerable on the IUCN Red List and in Appendix 3 of CITES. Because of these protections, Alligator Snapping Turtles cannot be collected within the state for any purpose without a special collecting permit. Fishing via trot lines continues to increase adult mortality, as does damage from boat propellers. Reduction of adult mortality and maintenance of nesting sites along major rivers and creeks are the keys to long-term management of this species in Alabama. No head-starting or ranching operations have been attempted, but these might be viable conservation options.

Taxonomy Thomas et al. (2014) provide convincing evidence for recognizing specimens from this genus in the Suwannee drainage of Florida and Georgia as a distinct species, *M. suwanniensis.* Based on mitochondrial data, they also elevate specimens from the Apalachicola and Choctawhatchee drainages to species status, assigning them to *M. apalachicolae.* Unfortunately, their diagnoses for *M. apalachicolae* and *M. temminckii* (from western drainages) are identical, with the exception of the median suture of the pygal (present in *M. apalachicolae* and absent in *M. temminckii*). Similarly, a morphometric analysis of skulls found no clear distinction between specimens from the Apalachicola/Choctawhatchee Rivers and drainages located further to the west (Murray et al. 2014). Therefore, if *M. apalachicolae* is to remain a distinct species, we find no way to distinguish it from *M. temminckii* in our keys. In this situation, we recognize a single species, *M. temminckii,* containing the eastern (Apalachicola and Choctawhatchee drainages) and western (Yellow to Trinity Rivers) mitochondrial lineages described first by Roman et al. (1999) and modified by Echelle et al. (2010) and Thomas et al. (2014). Our delimitation of *M. temminckii* is monophyletic with respect to *M. suwanniensis,* and these two taxa can be diagnosed from morphological features. We retain the common name Alligator Snapping Turtle for *M. temminckii* and note that previous authors have placed these turtles in the genera *Chelonura, Macroclemys,* and *Testudo.*

Mud and Musk Turtles

Family Kinosternidae

This family includes four genera with twenty-six species and ranges throughout most of the eastern and southern United States southward to northern South America. These turtles are relatively small and typically have drab-colored carapaces. Kinosternids spend a considerable amount of time on land either because their wetland habitat dries or because of periods of quiescence spent buried on land due to extreme heat or extreme cold. All forms have anatomical features associated with kinesis of the plastron, and in some forms the anterior and posterior portions of the plastron can be shut, protecting these turtles from attacks by predators or from extreme changes in weather. On the bridge, where the plastron and carapace come together, kinosternids have an opening to a gland that secretes a foul-smelling lipid that is presumed to ward off predators. This family is most closely related to the family Dermatemydidae of Central America (Krenz et al. 2005) and appears to have diverged as a distinct lineage 50 million years ago (Near et al. 2005). Two genera of this family are represented in North America and both occur in Alabama.

Key to the Genera of Kinosternidae of Alabama

1a Pectoral scutes triangular.

Genus ***Kinosternon*****—American Mud Turtles . . . . page 107.**

1b Pectoral scutes squarish or quadrangular.

Genus ***Sternotherus*****—Musk Turtles . . . . page 120.**

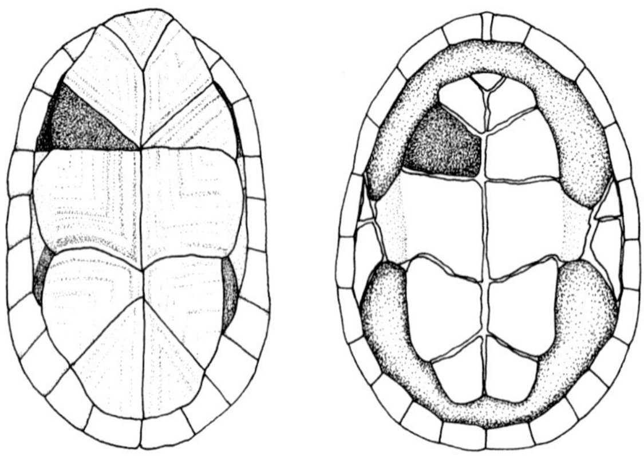

From left to right:

Triangular shape of abdominal scute of Mud Turtle

Squarish shape of abdominal scute of Musk Turtle

American Mud Turtles

Genus *Kinosternon* (Spix, 1824)

This widespread genus contains nineteen species that collectively span the geographic range of the entire family. The genus is distinguished by a triangular pectoral scute that creates a K-shaped pattern across the anterior portion of the plastron. This pattern also creates a hinge that allows some species to close the plastron. These features distinguish this genus from its closest living relative, the genus *Sternotherus,* which has a square or rectangular pectoral scute and is incapable of closing the plastron. Four species occur within the United States, two of which are found in Alabama.

Key to the Species of *Kinosternon* of Alabama

1a Three keels on carapace, highlighted by faint light stripes.

Kinosternon baurii—**Striped Mud Turtle . . . . page 109.**

1b Carapace uniform brown, lacking keels and light stripes.

Kinosternon subrubrum—**Eastern Mud Turtle . . . . page 117.**

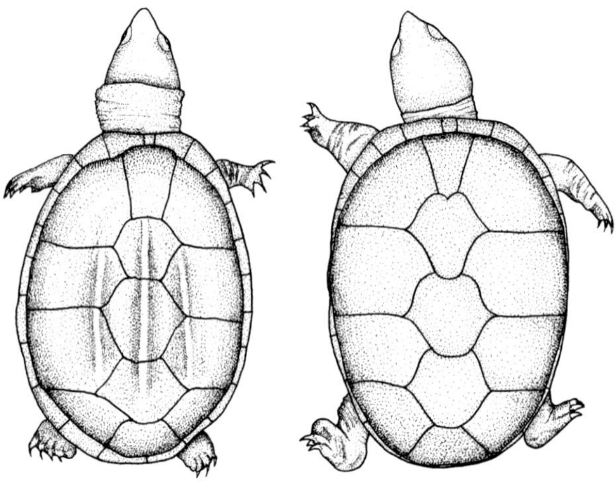

From left to right:

Carapace of Striped Mud Turtle with three keels

Carapace of Eastern Mud Turtle without keels

Striped Mud Turtle

Kinosternon baurii (Garman, 1891)

Description Striped Mud Turtles are small in size, attaining a maximum carapace length of around 4.5 inches (120 mm). The carapace is smooth, dome-shaped, and tallest at the posterior end; juveniles possess a weak mid-dorsal keel. The ground color of the carapace is brown to black, with some individuals being uniform and others possessing three faded tan stripes. The vertebrals may be depressed, forming a mid-dorsal groove. The first vertebral scute widens anteriorly to meet the seam between the first and second marginal scutes. The pectoral laminae of the plastron are triangular in shape, creating a forelobe of the plastron that is movable. The ground color of the soft parts is dark brown. Typically, the head and neck are mottled with dull yellow or yellowish brown that form a stripe above the eye and a stripe from the back of the mouth extending posteriorly along the neck. However, these stripes may become so heavily dissected that only a stripe from the external nares to the eye remains. The upper jaw is not strongly hooked at its anterior tip. Adult males possess a well-developed blunt spine or nail at the end of the tail as well as rough patches on the medial surfaces of the hind limbs. Tail length of mature males is longer

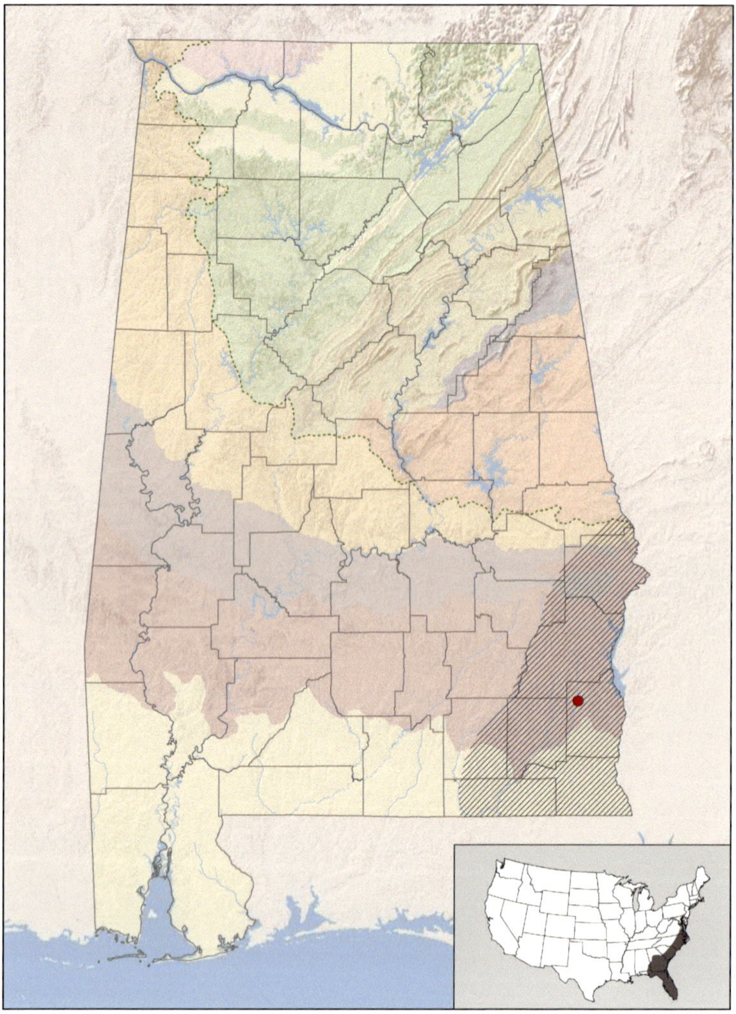

Right: Alabama and continental US distribution of the Striped Mud Turtle

Previous: Adult Striped Mud Turtle, Florida

and the base of the tail is noticeably more bulbous in males than in females. This species is identical in morphology to *K. subrubrum hippocrepis,* a subspecies that lacks the light stripes on the carapace found on some individuals of *K. baurii.*

Alabama Distribution This species was first recorded for Alabama by Sean Graham in 2010 based upon a specimen from the upper reaches of the Choctawhatchee River. We use this individual to document the species from the state and to infer its distribution. However, two additional specimens, one from Tuskegee National Forest in Macon County and the other from a creek in Lowndes County, await molecular analyses that may expand the influence of this species within Alabama.

Habits Striped Mud Turtles occupy slow-moving or stagnant waters, usually with abundant aquatic vegetation. Cypress swamps and blackwater streams are favored by the species. These animals are omnivorous, consuming aquatic vegetation, insects, mollusks, and carrion. Adults mature in five to six years. Mating may occur throughout the season of activity (Iverson 1977), with eggs being laid at any time during the season of activity. Nests are excavated in loose soil or decaying vegetation and contain up to five white, elliptical, hard-shelled eggs. Eggs produced late in the year enter diapause, with development slowing during winter and hatchlings emerging the following spring. Eggs may be fertilized with sperm that is stored for over a year (Nijs and Navez 1990).

Conservation and Management No more than two specimens of this species may be possessed at any point in time and these must be for personal use. Harvesting of Striped Mud Turtles for profit is prohibited. Members of the species thrive in a variety of wetlands, including those created by humans, and so special protection for the species within the state is not likely to be necessary. However, the species is rare and needs stronger documentation to determine its precise geographic distribution.

Taxonomy This species demonstrates no subspecific variation. However, specimens from Georgia and South Carolina become difficult to distinguish from *K. subrubrum* (Lamb and Lovich 1990) because the distinctive color patterns of *K. baurii* populations in Florida (three stripes on carapace and two bold yellow stripes on side of head) become muted in northern populations such that only a light stripe from the external nares to the eye remains. Morphologically, this species is indistinguishable from *K. subrubrum hippocrepis* (Lamb and Lovich 1990; Lovich and Lamb 1995), a feature that may indicate the two are conspecific.

Eastern Mud Turtles

Kinosternon subrubrum (Lacépède, 1788)

Taxonomy This common species is found throughout much of eastern North America. Three subspecies are recognized, two of which occur in Alabama. Previous authors have placed this species in the genera *Cinosternum, Cistuda, Swanka,* and *Testudo.*

Key to the Subspecies of *Kinosternon subrubrum* of Alabama

1a Side of head with two conspicuous light stripes; anteriormost vertebral scute wide anteriorly, meeting the seam between the first and second marginal scutes.

***Kinosternon subrubrum hippocrepis*—Mississippi Mud Turtle . . . . page 114.**

1b Side of head variously marked but lacking two conspicuous light stripes; anteriormost vertebral scute narrow anteriorly, meeting marginals along first marginal scute.

***Kinosternon subrubrum subrubrum*—Eastern Mud Turtle . . . . page 117.**

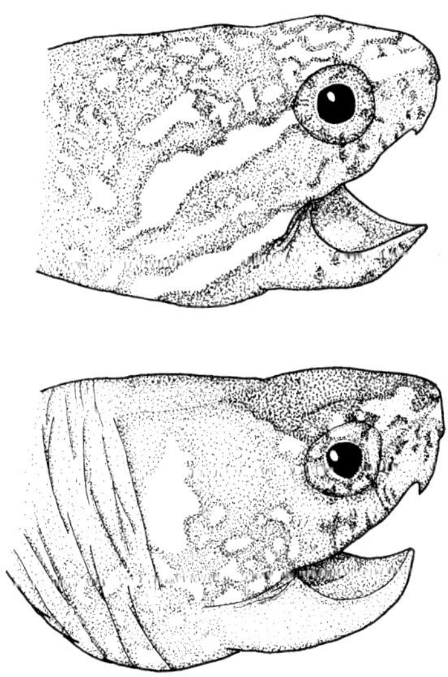

From top to bottom:

Lateral view of head and neck of Mississippi Mud Turtle with stripes

Lateral view of head and neck of Eastern Mud Turtle without stripes

Mississippi Mud Turtle

Kinosternon subrubrum hippocrepis (Gray, 1855)

Description Mississippi Mud Turtles are small in size, attaining a maximum carapace length of around 4.5 inches (120 mm). The carapace is dark brown to black, smooth, dome-shaped, and tallest at the posterior end; juveniles have a mid-dorsal keel and a pair of weak dorsolateral keels. The first vertebral scute is wide, joining the marginals near the suture between the first and second marginals. The pectoral laminae of the plastron are triangular in shape, creating a forelobe of the plastron that is movable. The ground color of the soft parts is dark brown. The head and neck are mottled with dull yellow or yellowish brown that form one or two diffuse stripes extending from behind the eye along the neck. The upper jaw is strongly hooked at its anterior tip. Adult males possess a well-developed blunt spine or nail at the end of the tail, as well as rough patches on the medial surfaces of the hind limbs. Tail length is longer and the base of the tail is noticeably more bulbous in males than in females. The plastron of hatchlings is reddish orange to yellowish with a dark central figure. Eastern Mud Turtles, the most similar taxon, lack evidence of stripes behind the eye and have a narrow first vertebral scute, meeting the marginals

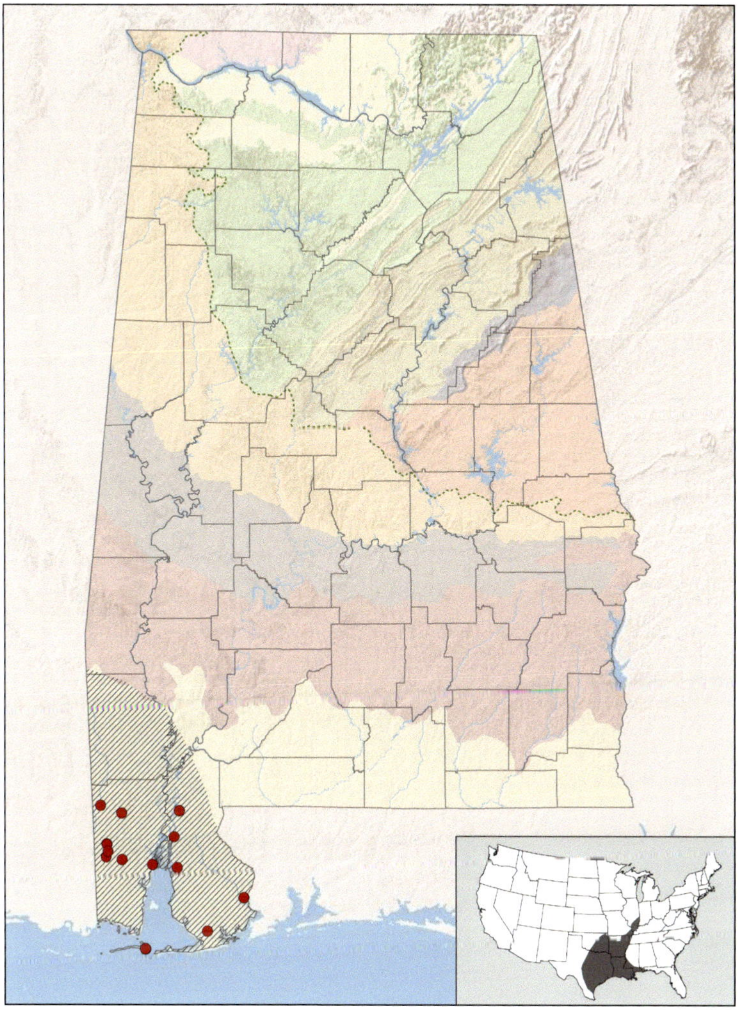

Left: Alabama and continental US distribution of the Mississippi Mud Turtle

Opposite: Adult Mississippi Mud Turtle, Mobile County, AL

near the center of the first marginal scute. Mississippi Mud Turtles also lack the three carapacial stripes of Striped Mud Turtles, a second similar taxon.

Alabama Distribution Mississippi Mud Turtle populations inhabiting the Baldwin County area are intergradient between this subspecies and *K. s. subrubrum*; those from Mobile County appear to be *K. s. hippocrepis.*

Habits Mississippi Mud Turtles are abundant in extreme southwestern Alabama, where they may be found in ponds, lakes, swamps, marshes, and flooded roadside ditches. The subspecies usually is found crawling about on the bottom of stagnant waters where aquatic vegetation is lush. During winter, Mississippi Mud Turtles may move from aquatic sites to upland terrestrial sites where they bury themselves under loose soil and leaf litter.

The nest of this subspecies is dug in the ground or in piles of organic matter, including abandoned alligator nests. Most nests contain two to five eggs that are hard-shelled and quite resistant to desiccation.

Conservation and Management As with all of Alabama's turtles, Mississippi Mud Turtles cannot be harvested for commercial purposes. Two individuals per day can be possessed, but only for personal uses. This subspecies does well in areas altered by human activities. It is likely to have increased in Alabama in association with creation of farm ponds and other managed wetlands. For these reasons, no special conservation or management activities are necessary to maintain Mississippi Mud Turtles in the state of Alabama.

Eastern Mud Turtle

Kinosternon subrubrum subrubrum (Lacépède, 1788)

Description Eastern Mud Turtles are small in size, attaining a maximum carapace length of around 4.5 inches (120 mm). The carapace is plain brown to black, smooth, dome-shaped, and tallest at the posterior end; juveniles have a mid-dorsal keel and a pair of weak dorsolateral keels. The first vertebral scute is narrow, joining the marginals near the center of the first marginal. The pectoral laminae of the plastron are triangular in shape, creating a forelobe of the plastron that is movable. The ground color of the soft parts is dark brown. Typically, the head and neck are mottled with dull yellow or yellowish brown, but these markings do not form bold striping. The upper jaw is strongly hooked at its anterior tip. Adult males possess a well-developed blunt spine or nail at the end of the tail as well as rough patches on the medial surfaces of the hind limbs. Tail length is longer, and the base of the tail is noticeably more bulbous in males than in females. The plastron of hatchlings is reddish orange to yellowish with a dark central figure. This species is larger in northern populations than in southern ones (Ashton and Feldman 2003), a pattern that appears to be true for Alabama specimens above and below the Fall Line (Mount 1975).

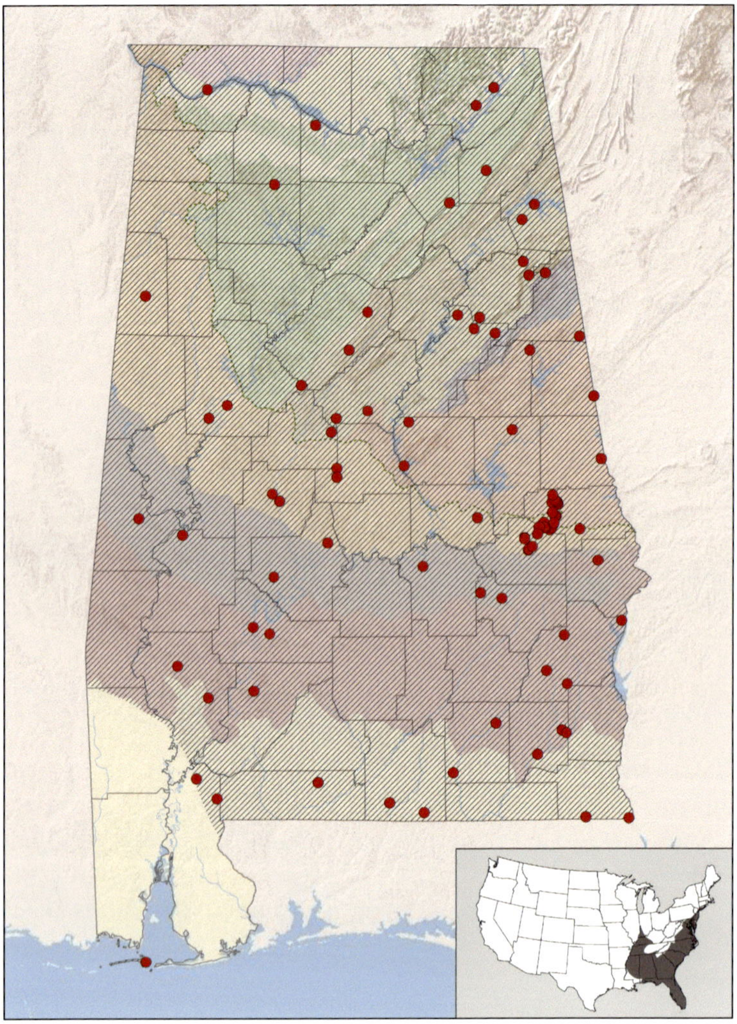

Right: Alabama and continental US distribution of the Eastern Mud Turtle

Previous: Adult Eastern Mud Turtle, Elmore County, AL

Eastern Mud Turtles are difficult to distinguish from *K. subrubrum hippocrepis* and *K. baurii,* but both of these have light head stripes and a first vertebral scute that is widest at its anterior, meeting the seam between the first and second marginal scutes.

Alabama Distribution This subspecies is found statewide, except in Mobile and Baldwin Counties, where it intergrades with and is replaced by *K. s. hippocrepis.*

Habits The Eastern Mud Turtle is abundant in Alabama and may be found in ponds, lakes, swamps, marshes, and flooded roadside ditches. About the only aquatic habitats likely to be shunned are free-flowing creeks and rivers.

The Eastern Mud Turtle frequently wanders about on land, is more active at night than during the day, and seldom basks. In its aquatic habitats, it is usually seen crawling about on the bottom. During winter, Eastern Mud Turtles move from aquatic sites to upland terrestrial sites where they create shallow (0.8 inches [2 cm]) hibernacula by burying themselves under loose soil and leaf litter; here they generally are inactive for an average of about 100 days (Steen et al. 2007). Individuals may emerge on warm winter nights and forage in rain-flooded depressions where they may kill and consume winter-breeding frogs (*Pseudacris ornata* and *Lithobates sphenocephalus*; Mount 1975).

The nest of this species is dug in the ground or in piles of organic matter, including abandoned beaver dams and old alligator nests. Most Eastern Mud Turtle nests contain two to five eggs that are hard-shelled and strongly resistant to desiccation.

Conservation and Management This subspecies is well suited to human activities and is likely to have increased in abundance in Alabama in association with creation of farm ponds and other managed wetlands. Nevertheless, this species cannot be harvested for profit. Instead, a maximum of two individuals may be possessed per day for personal use. Eastern Mud Turtles can survive in urban settings, especially in parks and golf courses with managed ponds. For these reasons, no special conservation or management activities are necessary to maintain the species in the state of Alabama.

Musk Turtles

Genus *Sternotherus* (Gray, 1825)

We recognize five species within this genus, all of which occur within Alabama. These turtles inhabit streams, ponds, and lakes, generally along the aquatic margin and frequently in flowing waters. They are most similar to the genus *Kinosternon*, their closest living relative. However, the two genera are easily differentiated by the presence of a square or rectangular pectoral scute and immobile plastron in *Sternotherus* (triangular pectoral and mobile plastron in *Kinosternon*).

Key to the Species of *Sternotherus* of Alabama

1a Barbels on chin and neck; side of head usually with two conspicuous light stripes.

Sternotherus odoratus—**Eastern Musk Turtle . . . . page 123.**

1b Barbels on chin only; side of head variously marked, but lacking two conspicuous light stripes. Go to **2.**

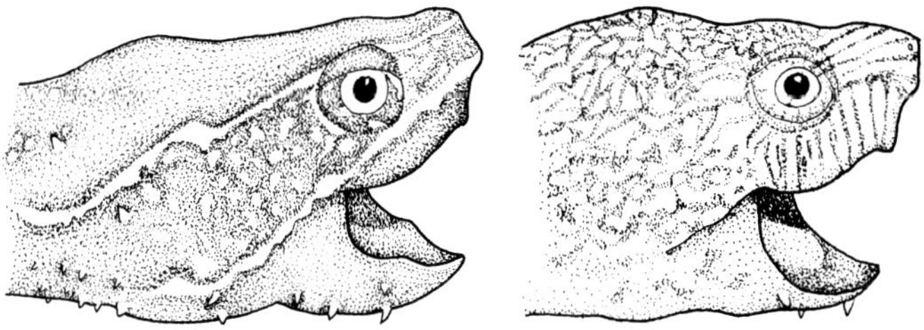

From left to right:

Lateral view of head and neck of Eastern Musk Turtle with barbels on chin and neck

Lateral view of head and neck of Flattened Musk Turtle with barbels on chin only

2a Plastron lacking a gular scute; carapace high-keeled, about as tall as it is wide.

Sternotherus carinatus—**Razor-backed Musk Turtle . . . . page 126.**

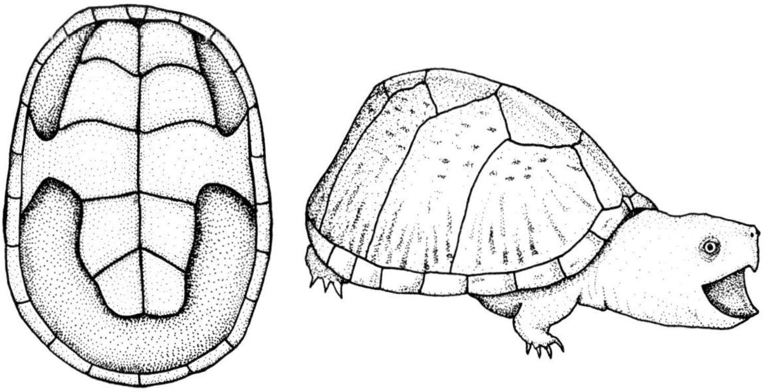

From left to right:

Plastron of Razor-backed Musk Turtle without gular scute

High-keeled carapace of Razor-backed Musk Turtle

2b Plastron with gular scute; carapace wider than it is tall. Go to **3**.

From left to right:

Plastron of Loggerhead Musk Turtle with gular scute

Carapace of Flattened Musk Turtle with low keel

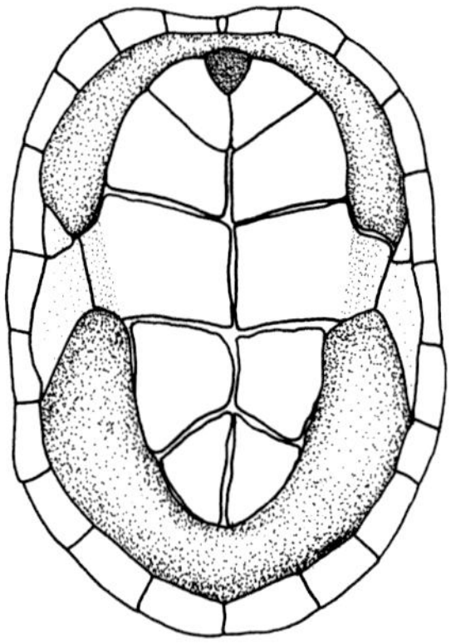

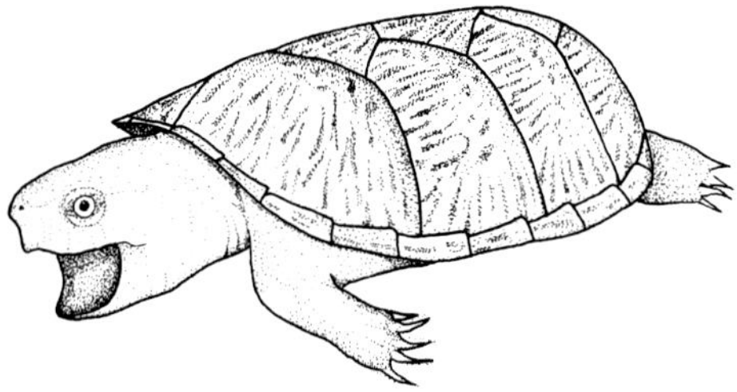

3a Shell strongly flattened; top of head greenish with a dark reticulum; top of snout spotted or blotched.

***Sternotherus depressus*—Flattened Musk Turtle . . . . page 129.**

3b Shell not strongly flattened; head and snout not pigmented as above. Go to **4**.

4a Distinct dark spots posterior to eye; neck, at most, weakly striped.

***Sternotherus minor*—Loggerhead Musk Turtle . . . . page 133.**

4b Curved dark stripes posterior to eye; neck boldly marked with light and dark stripes.

***Sternotherus peltifer*—Stripe-necked Musk Turtle . . . . page 137.**

From left to right:

Lateral view of head and neck of Loggerhead Musk Turtle with dark spots

Lateral view of head and neck of Stripe-necked Musk Turtle with dark stripes

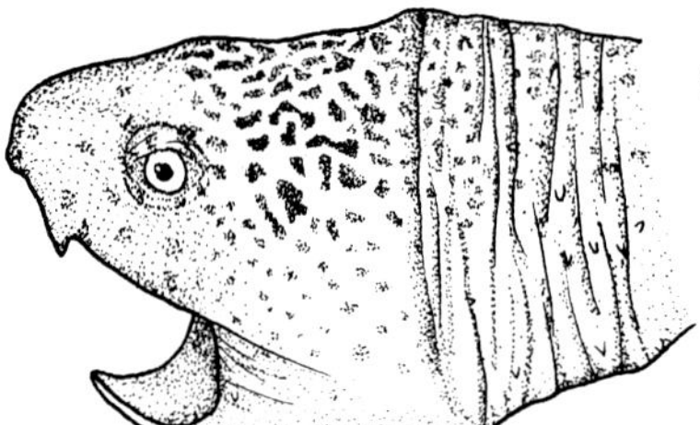

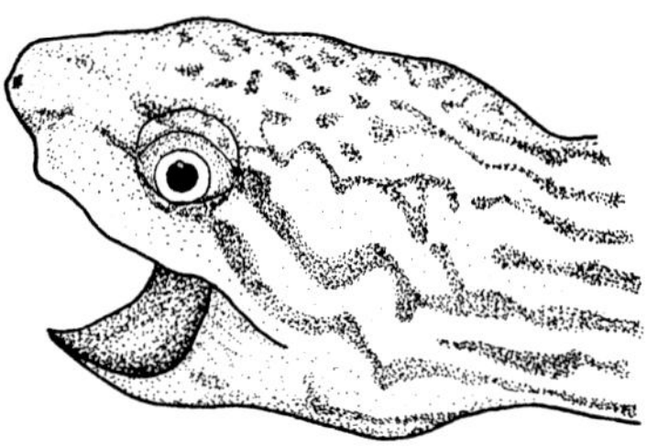

Eastern Musk Turtle (Stinkpot)

Sternotherus odoratus (Latreille, 1801)

Description This is a moderately small, aquatic turtle, attaining a maximum carapace length of about 5.5 inches (135 mm). The scutes of the carapace do not overlap except in young animals. The plastron is relatively small, has a single gular at its anterior terminus, and has an anterior lobe that is slightly movable. The pectoral scute is rectangular or quadrangular in shape. Barbels are present on the chin and neck. In ground color, the carapace is light olive brown to dark brown, often with radiating dark streaks. The carapace of juveniles has a median keel. Key field characters are the two bold light stripes (rarely lacking in adults), one extending from the tip of the snout, through the eye, to the neck, and the other extending from the tip of the snout, along the upper lip, to the neck. The plastron of adults is brownish to yellowish. Hatchlings have a plastron that is mottled or marbled, marginals that have light markings, and a head with conspicuous stripes. Adult males have a blunt spine or nail at the end of the tail and rough patches on the insides of the hind legs.

Alabama Distribution This species is found throughout the state.

Above: Adult Eastern Musk Turtle with strong head striping, Madison County, AL

Previous: Adult Eastern Musk Turtle with weak head striping, Jackson County, AL

Habits In Alabama, the Eastern Musk Turtle inhabits ponds, lakes, sloughs, and other still- or sluggish-water habitats. Extremely aquatic, it spends most of its time crawling about on the bottom and seldom basks.

A small clutch of elongate, hard-shelled eggs, usually three to five in number, is laid in a shallow excavation dug in a mound of loose soil, or in a pile of rotting organic debris such as a decaying log or stump. The sex of offspring from these eggs is determined by the temperature of the nest, with females produced from warmer nests and males produced from cooler ones (Janzen and Paukstis 1991). The food of *S. odoratus* includes carrion, a wide variety of small aquatic invertebrates, and small amounts of vegetation. Freshly captured individuals often secrete a foul-smelling musk and will bite if given the opportunity.

Conservation and Management Eastern Musk Turtles can be abundant in appropriate habitat, but the species is susceptible to population declines if harvested, as are all turtles. Therefore, it is illegal to harvest them for profit. However, up to two individuals per day may be possessed for personal use. Current best management practices for

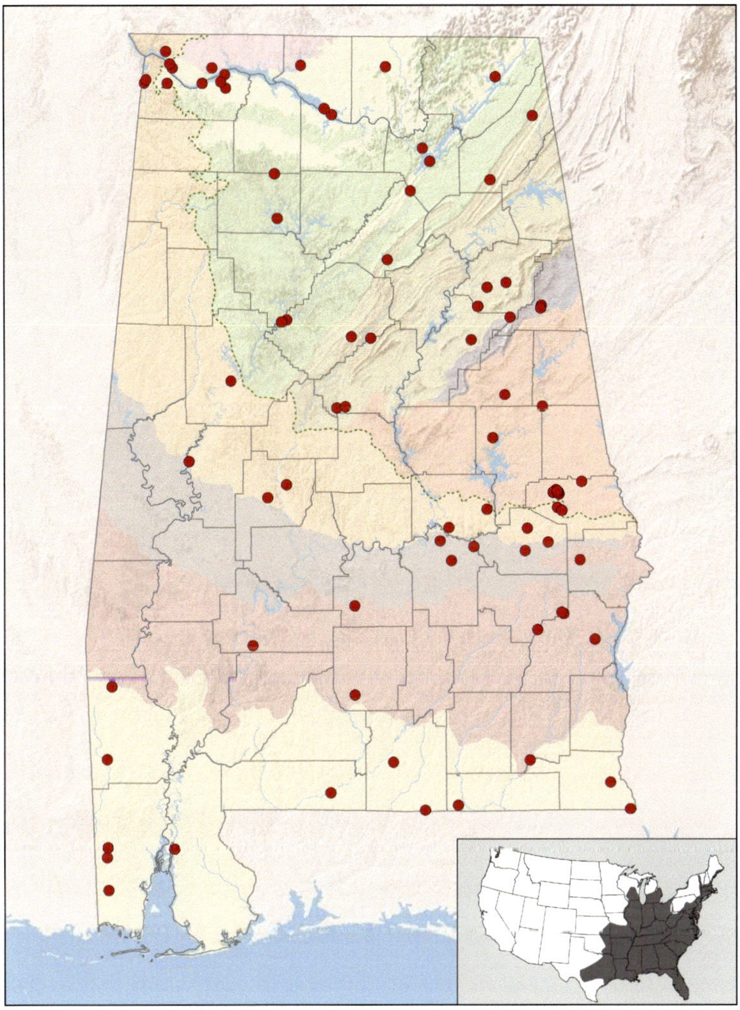

Alabama and continental US distribution of the Eastern Musk Turtle

streamside management zones are likely to be sufficient for protecting habitat for this species. However, because members of this species, especially the females, cross roads frequently (Steen et al. 2006), wildlife fencing and designed underpasses in areas of heavy traffic may be necessary to maintain populations in urban settings.

Taxonomy No subspecific variation has been described for this species. Previous authors have placed it in the genera *Aromochelys, Cinosternum, Kinosternon, Kinosternum, Ozotheca, Staurotypus, Sternothaerus, Terrapene,* and *Testudo.*

Razor-backed Musk Turtle

Sternotherus carinatus (Gray, 1855)

Description Razor-backed Musk Turtles are aquatic animals that are moderate in size, attaining a maximum carapace length of about 7 inches (180 mm) and possessing a carapace that is as tall as it is wide. The center of the carapace has a prominent keel. Unlike most other members of the genus in Alabama, the plastron is relatively wide, covering much of the venter. The anterior lobe of the plastron lacks gular scutes and this region is only slightly movable. The pectoral scutes are rhomboid in shape and barbels are present on the chin only. The dominant ground colors of the carapace are light brown to orange, with a series of thin dark streaks. The plastron is an immaculate light yellow. In ground color, the top of the head and snout are light brown to gray, with numerous, mostly rounded spots.

Alabama Distribution In Alabama this species is found only in relatively deep portions of the Escatawpa River in Washington and Mobile Counties.

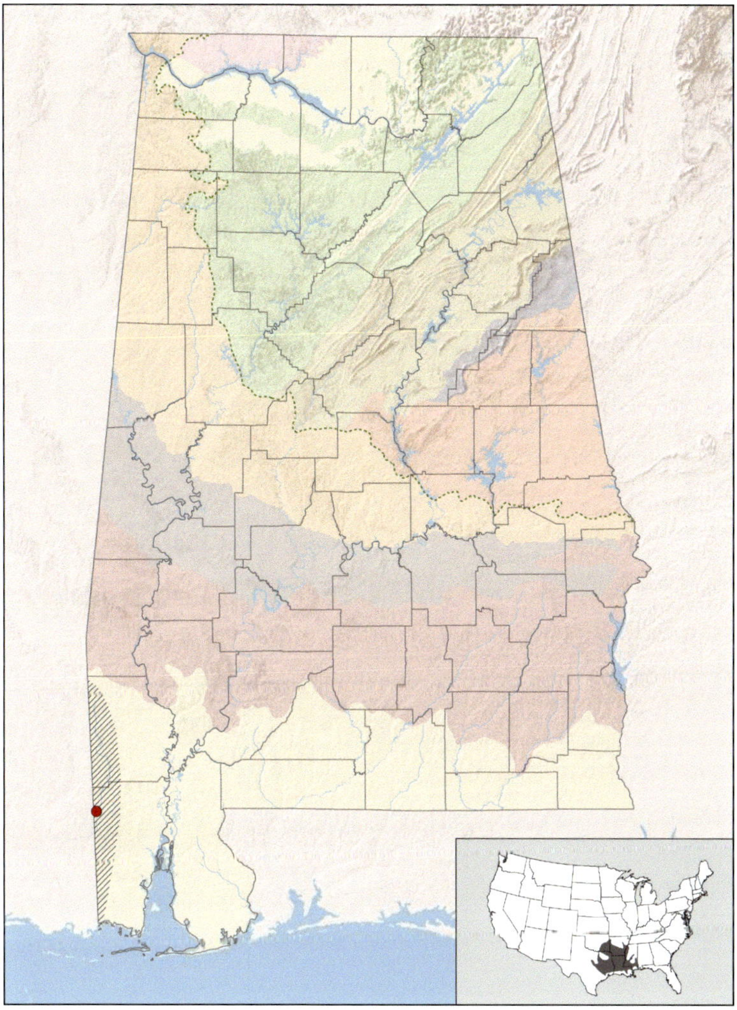

Left: Alabama and continental US distribution of the Razor-backed Musk Turtle

Opposite: Adult Razor-backed Musk Turtle, Mobile County, AL

Habits Unlike most members of the family, Razor-backed Musk Turtles are fond of basking. It is therefore most frequently observed on logs above water. When approached closely, these turtles will drop into the water, often from remarkable heights. Seasonal movements of this species are relatively restricted (Mahmoud 1969). Mating occurs in May and June and females lay about five eggs per clutch; the eggs are elliptical and have a hard white shell (Ernst and Lovich 2009). This species is carnivorous, consuming a wide variety of freshwater invertebrates (Ernst and Lovich 2009).

Conservation and Management Razor-backed Musk Turtles are documented for Alabama from only two specimens collected in 1999 (AUM 33725) and 2000 (AUM 36382). The species is ranked Priority 2 (High Conservation Concern) by ADCNR (Mirarchi et al. 2004) and is monitored by the Alabama Natural Heritage Program as critically imperiled because so few individuals may be present in the state. For this reason it is unlawful to possess these turtles without a special permit. In the 1960s and 1970s Auburn University field biologists visited the region in Alabama occupied by this species and yet did not observe it, which suggests that it is a relatively new immigrant to Alabama from the west. The species is common in Mississippi, and so Alabama is unlikely to play a large role in any conservation efforts that might be needed for these turtles. The SWAP (Alabama Department of Conservation and Natural Resources, Division of Wildlife and Freshwater Fisheries 2005) recommends comprehensive surveys for this species every five to ten years to monitor its status within the state. Maintenance of snags and streamside management zones for wildlife should be sufficient to provide appropriate habitat for this species.

Taxonomy No subspecies are currently recognized. Previous authors have placed this species in the genera *Aromochelys*, *Goniochelys*, and *Kinosternon*.

Flattened Musk Turtle

Sternotherus depressus (Tinkle and Webb, 1955)

Description Flattened Musk Turtles are highly aquatic and reach a maximum size of about 5 inches (125 mm). They have a carapace that is extremely flattened dorsoventrally. In ground color the carapace is light to dark brown with a series of dark brown slash marks or spots. The plastron is pinkish in young turtles and yellowish in adults. The top of the head is greenish yellow with a reticulum of dark markings, which often break up to form spots and blotches on top of the snout. Along the sides of the head and neck, light and dark markings align to create a weakly striped color pattern. The heads of old adults are not greatly enlarged. The shape of the carapace distinguishes this species from all other members of the genus.

Alabama Distribution This species is confined to Alabama, where it occurs in permanent streams of the Black Warrior River system above the Fall Line. Estridge (1970), based on color pattern, concluded that *S. depressus* hybridizes with *S. peltifer* in the Upper Cahaba River system and in a small portion of the Warrior River system along the Fall Line. But Ernst et al. (1988) concluded that two distinct taxa could be distinguished based on shell morphology and placed the Cahaba River specimens in *S. peltifer.* We consider the putative hybrids to belong to *S. peltifer.*

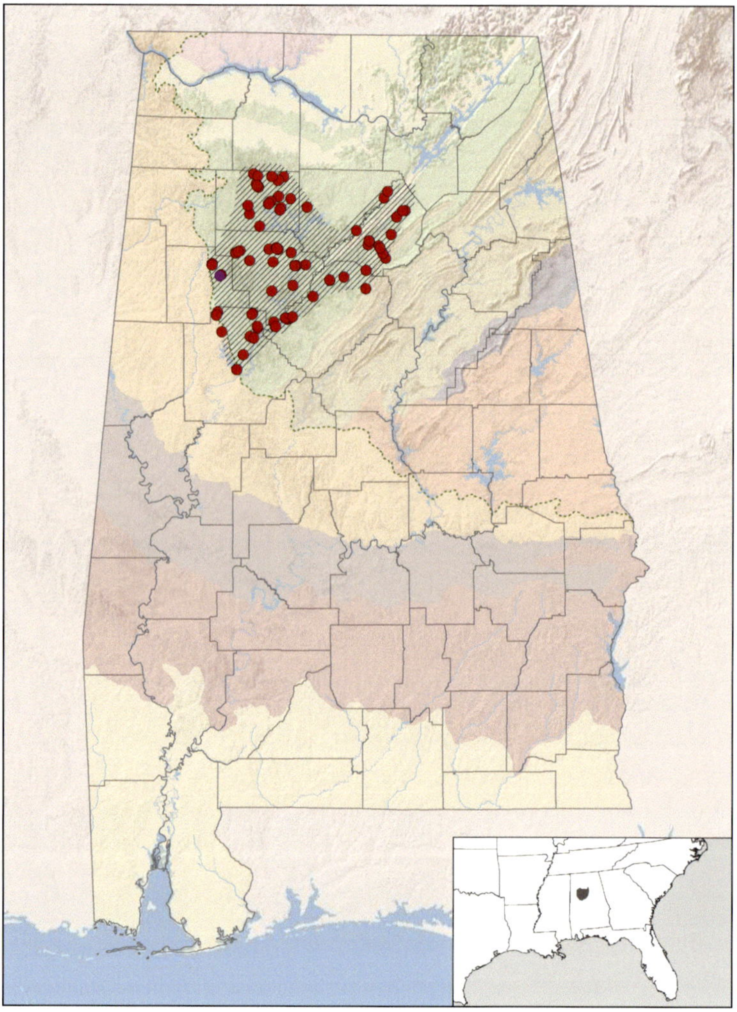

Right: Alabama distribution of the Flattened Musk Turtle; purple dot indicates hybrid with the Stripe-necked Musk Turtle

Previous: Adult Flattened Musk Turtle, Winston County, AL

Habits This species inhabits the bottom of relatively clear, rocky streams that are at least 1–3 feet (0.3–1 m) in depth. Watersheds of less than 8 square miles (21 km^2) generally are not inhabited. Deep impoundments with rocky shorelines are occupied but ponds are not. Active animals can be observed from April through October and their overwintering sites are unknown. The species has a strong tendency for nocturnal activity, especially in adults (Estridge 1970), and is strongly tied to water, moving from rock to rock when active. During daylight hours these turtles remain under large flat rocks where, like other members of their genus, individuals are capable of remaining submerged for many hours without surface breathing.

The food of *S. depressus* consists largely of mollusks, although other invertebrates may be eaten (Schnuelle 1997). Sites that are affected by surface runoff have turbid water that limits exposure of bedrock, reducing the availability of invertebrate prey. These features appear to limit juvenile Flattened Musk Turtles. Basking does occur in healthy members of this species, but this behavioral feature is especially common in sick individuals (Dodd et al. 1988).

Females produce eggs during June and July, laying one to three (usually two) eggs per clutch in up to two clutches per year (Close 1982, Bailey and Bailey 2002). Nests are placed along streams and under vegetation. Females are known to travel as little as 16.5 feet (5 m) and as much as 377 feet (115 m) from the stream to the nest site. These nests may be as little as 3.5 feet (1 m) and as much as 105 feet (32 m) above stream base flow (Bailey and Bailey 2002). The eggs are oblong and have hard but brittle shells. Incubation may take as little as forty-five days (incubated at 25° C in a lab; Dodd et al. [1988]), but may require seventy-six days in field settings (Bailey and Bailey 2002).

Conservation and Management The Flattened Musk Turtle is listed as Threatened by the USFWS (1987) and therefore federally protected under the ESA. It is ranked Priority 2 (High Conservation Concern) by ADCNR (Mirarchi et al. 2004) and, therefore, is protected by the state of Alabama. Finally, the species is listed as vulnerable by the IUCN. Population data document that Flattened Musk Turtles are affected by severe levels of habitat fragmentation (Dodd 1990) and episodes of increased mortality due to disease (Fonnesbeck and Dodd 2003). These factors appear to cause a gradual decline in capture rate, maximum body size, and proportion of females captured, even in the best remaining sites (Bailey and Guyer 1998). This has occurred despite federal listing. No recovery plan has been generated for the species. Clearly more needs to be done to protect Flattened Musk Turtles, and modification of timber and mining activities appear to be top priorities in effecting such conservation. In areas where logging and mining have been restricted, some streams have been restored to near-pristine conditions by decreasing sediment loads and increasing water clarity, but this has not necessarily increased numbers of turtles. Because extreme fragmentation separates populations to such a degree that recovery may be unacceptably slow, and because turtles can be reared in captivity fairly easily, head-starting might be necessary to increase breeding stocks in some areas. Additionally, studies of nesting habits are needed to determine whether nest sites are receiving sufficient protection. The SWAP (Alabama Department of Conservation and Natural Resources, Division of Wildlife and Freshwater Fisheries 2005) recommends comprehensive surveys for this species every five to ten years to monitor its status within the state.

Taxonomy No subspecies currently are recognized. These turtles were considered by Mount (1975) to belong to *Sternotherus minor*, a taxon that was also conceived to include *S. peltifer*. Walker et al. (1998) provided a phylogenetic tree in which *S. depressus* was shown to be a distinct monophyletic lineage. That study also showed *S. minor* and *S. peltifer* to be monophyletic sister taxa, but generated the surprising relationship of *S. depressus* being sister to *S. carinatus* + *S. odoratus*. This is surprising because of the strong similarity between *S. depressus* and *S. peltifer* in morphology and color pattern and because of the equally strong dissimilarities between *S. depressus* and *S. carinatus* or *S. odoratus*. Based on available data, we accept *S. depressus* as a distinct species, but remain convinced that the addition of color and morphological data to available sequence data will show this species to be sister to or nested within *S. peltifer*. A few previous publications have placed *S. depressus* in the genus *Kinosternon*.

Loggerhead Musk Turtle

Sternotherus minor (Agassiz, 1857)

Description This is a moderately small aquatic turtle attaining a maximum carapace length of about 5 inches (130 mm). The scutes of the carapace overlap one another. The plastron is relatively small, and its anterior lobe is slightly movable. The pectoral scute is rectangular or quadrangular in shape and one gular scute is present at the anterior end. Barbels are present on the chin only. The carapace has three distinct keels, which are most pronounced in juveniles and often absent in old adults. The carapace is brown, with dark spots or streaks that become obscure in old adults. For specimens from the Chattahoochee, Choctawhatchee, and Pea Rivers, including individuals from Big Creek and Cowarts Creek of the Chipola River, the top of the head and snout are light brown to gray and have numerous rounded dark spots that descend to the region behind the eye. The back of the head and sides of the neck have similar markings that are less regular in appearance. But on individuals from the Conecuh and Yellow Rivers to the eastern edge of Mobile Bay, head and neck patterns become variable. Spots on the head and neck may become faded or absent, and spots may align as faint light and dark neck stripes. Mount (1975)

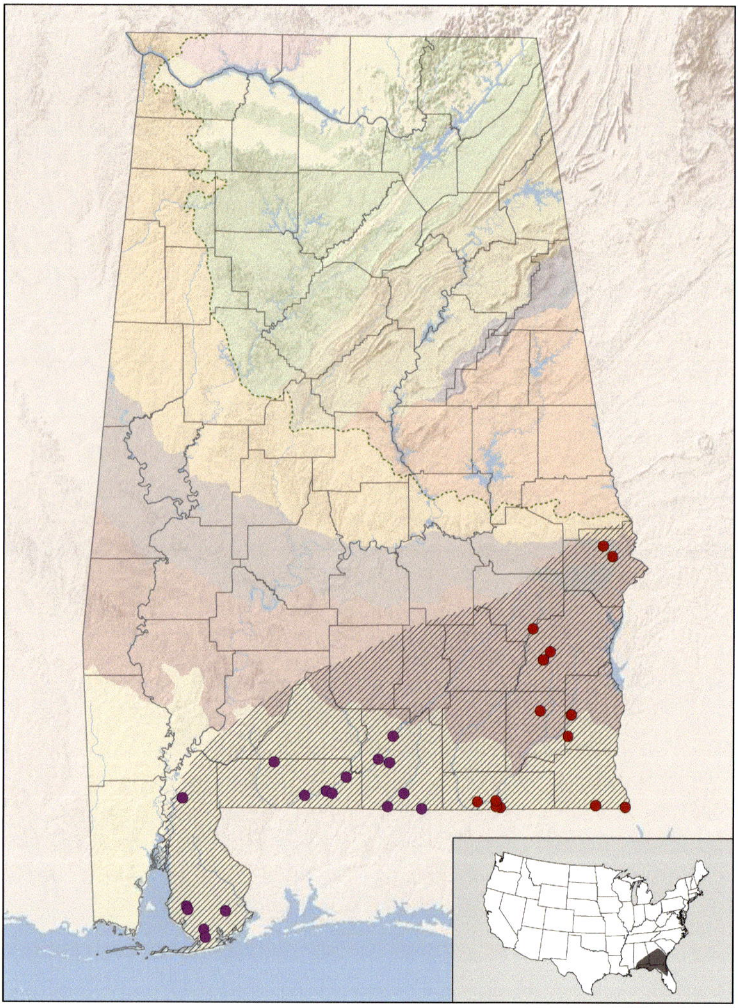

Right: Alabama and continental US distribution of the Loggerhead Musk Turtle with typical color pattern (red dots) and modified color pattern (purple dots)

Previous: Adult Loggerhead Musk Turtle, Covington County, AL

interpreted these to be intergradient between *S. m. minor* and *S. m. peltifer*. However, emerging genetic data suggest these may represent subspecific variation within our concept of *S. minor*. The plastron is pink in all young animals, becoming yellow with age and the head becomes greatly enlarged in old adults. In males, the side of the tail has a series of spots. Adult males also have a blunt spine or nail at the end of tail and rough patches on the insides of the hind legs. This species is quite similar in appearance to *S. peltifer*, but that species has dark worm-like stripes behind the eye that extend on to the neck as bold light and dark stripes; additionally, the lateral surface of the tail is striped in male *S. peltifer.*

Alabama Distribution This species is found from drainages east of Mobile Bay to the Apalachicola River drainage in Houston, Henry, and Barbour Counties. Individuals from the Fish and Perdido Rivers of Baldwin County are referable to this species. Those from the Dog and Tensaw Rivers of Mobile County and the Mobile Bay drainage area north of Mobile County are referable to *S. peltifer.*

Habits Although the Loggerhead Musk Turtle is sighted occasionally in farm ponds and other isolated habitats with standing water, it is primarily an inhabitant of creeks and rivers with flowing waters. It is frequently encountered early in the morning, crawling on the bottom around logs, rocks, and masses of vegetation. Insects and mollusks are the principal food items. The large heads and expanded crushing surfaces of the jaws in adults are presumably adaptations to feeding on mollusks.

Sternotherus minor is seldom seen basking in the open in the manner of other river and pond turtles but will often climb onto a leafy branch of a tree protruding from the water, sometimes to surprising heights. Otherwise, it seldom leaves aquatic habitats except to nest. Females reach sexual maturity at age nine (3 inches [80 mm] carapace length), while males mature at five to six years (2 inches [55 mm] carapace length; Cox et al. 1991). Up to four eggs are produced in each clutch and females produce two to three clutches per year (Cox and Marion 1978). This species produces hard-shelled eggs. Nests are placed on high banks near water or, where the landscape slopes gently from water's edge, far from water. The sex of offspring from these eggs is determined by the temperature of the nest, with females produced from warmer nests and males from cooler nests (Janzen and Paukstis 1991).

Conservation and Management The Loggerhead Musk Turtle is abundant in appropriate habitats. Nevertheless, it is illegal to capture these animals for profit because, like all turtles, the species is susceptible to population decline from harvest pressure. The effects of urbanization on streams may enhance habitat for this species (Barrett and Guyer 2008). Current best management practices for streamside management zones are likely to be sufficient for protecting its habitat.

Taxonomy Traditionally, two subspecies of this turtle have been recognized, both of which occur in Alabama (Mount 1975). However, the two differ in color pattern, largely occupy different river systems, and differ in growth trajectories, features that led Ernst et al. (1988) to recommend that the two subspecies be recognized as full species. Walker and Avise (1998), based on examination of the mitochondrial genome, showed the two subspecies to be reciprocally monophyletic sister taxa with a high degree of differentiation between the two lineages. Because the two taxa can be distinguished from color characters and occupy separate geographic regions of the state, we elevate *Sternotherus minor peltifer* to species status and restrict *Sternotherus minor* to those forms previously contained in *Sternotherus minor minor.* A few previous authors have placed this species in the genera *Goniochelys* and *Kinosternon.*

Stripe-necked Musk Turtle

Sternotherus peltifer (Smith and Glass, 1947)

Description This is a moderately small aquatic turtle attaining a maximum carapace length of about 4.5 inches (120 mm). The scutes of the carapace overlap one another. The plastron is relatively small, and its anterior lobe is slightly movable. The pectoral scute is rectangular or quadrangular in shape and one gular scute is present at the anterior end. Barbels are present on the chin only. The carapace has a single, mid-dorsal keel that is most pronounced in juveniles and often absent in old adults. The carapace is brown, with dark spots or streaks that become obscure in old adults. The top of the head and snout is light brown to gray and has numerous rounded dark spots that coalesce to form bold light and dark neck stripes starting at the back of the head and behind the eyes. These stripes are faded in individuals from the west side of Mobile Bay, but the postorbital stripes are still visible. The plastron is pink in young, becoming yellow with age, and the head becomes greatly enlarged in old adults. In males, the side of the tail has a series of dark lateral stripes. Adult males also have a blunt spine or nail at the end of the tail and rough patches on the insides of the hind legs. This species is quite similar in appearance to *S. minor*, but that species has distinct dark spots on the head immediately behind the

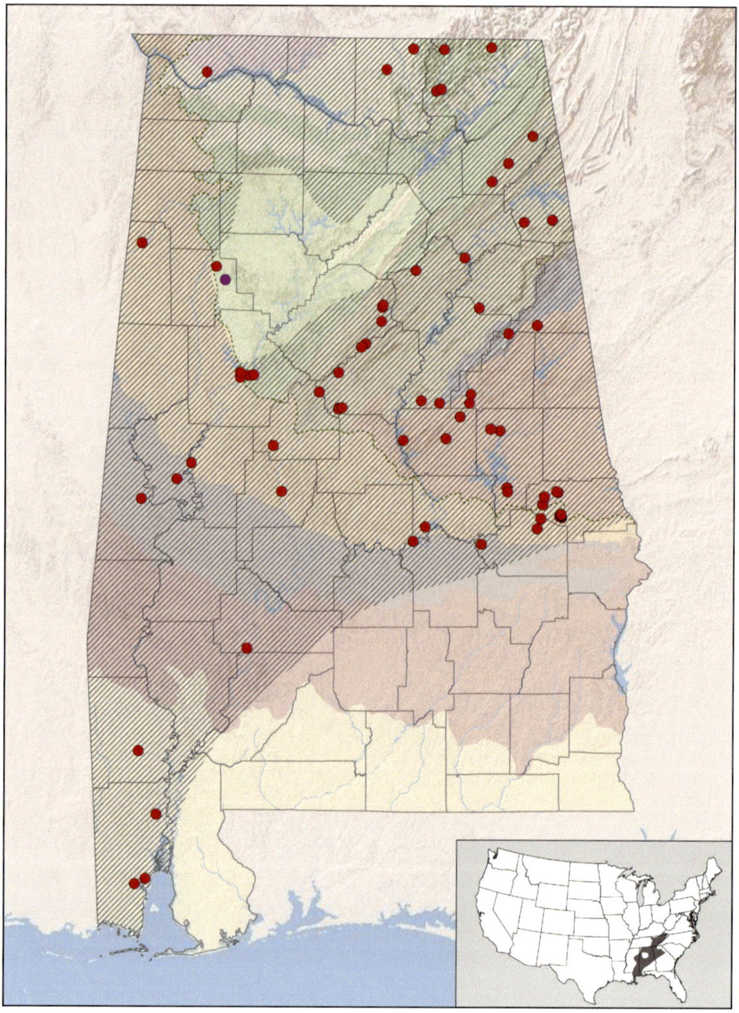

Right: Alabama and continental US distribution of the Stripe-necked Musk Turtle; purple dot indicates hybrid with the Flattened Musk Turtle

Previous: Adult Stripe-necked Musk Turtle, Alabama

eyes and lacks bold light and dark stripes on the neck, although faint stripes may be present; additionally, the lateral surface of the tail is spotted in male *S. minor*.

Alabama Distribution This species is found in the Tennessee River system, the Escatawpa River, and the Mobile Bay drainage, with the exception of the Fish River (occupied by *S. minor*) and the Black Warrior River above the Fall Line (occupied by *S. depressus*).

Habits Although the Stripe-necked Musk Turtle turns up occasionally in farm ponds, it is primarily an inhabitant of creeks and rivers and their associated aquatic environments. It is frequently encountered early in the morning, crawling on the bottom around logs, rocks, and masses of vegetation. Insects, mollusks, carrion, and algae are the principal food items. *Sternotherus peltifer* is seldom seen basking in the open but will often climb onto riparian vegetation protruding from the water. Otherwise, it seldom leaves aquatic habitats except to nest. Females lay up to four hard-shelled eggs in each clutch and may produce two to three clutches per year. Nests are placed on high banks near to water or, where the landscape slopes gently from water's edge, a considerable distance from water.

Conservation and Management Stripe-necked Musk Turtles reach exceptional densities in rocky areas of large rivers and streams. Nevertheless, it is illegal to capture this species for profit, although up to two individuals per day may be possessed for personal use. Current best management practices for streamside management zones are likely to be sufficient for protecting habitat for this species. Heavy sediment loads are likely to make some areas uninhabitable by this species, and management efforts to minimize such sedimentation are likely to benefit it.

Taxonomy Traditionally, this taxon was considered to be a subspecies of *S. minor* (Mount 1975). However, the two differ in color pattern, largely occupy different river systems, and differ in growth trajectories, features that led Ernst et al. (1988) to recommend that the two subspecies be recognized as full species. Walker and Avise (1998), based on examination of the mitochondrial genome, showed *S. m. minor* and *S. m. peltifer* to be reciprocally monophyletic sister taxa with a high degree of differentiation. Because the two taxa can be distinguished from color characters and occupy separate geographic regions of the state, we elevate *S. m. peltifer* to species status and restrict *S. minor* to those forms formerly contained in *S. m. minor.* A few previous authors have placed this species in the genera *Goniochelys* and *Kinosternon.*

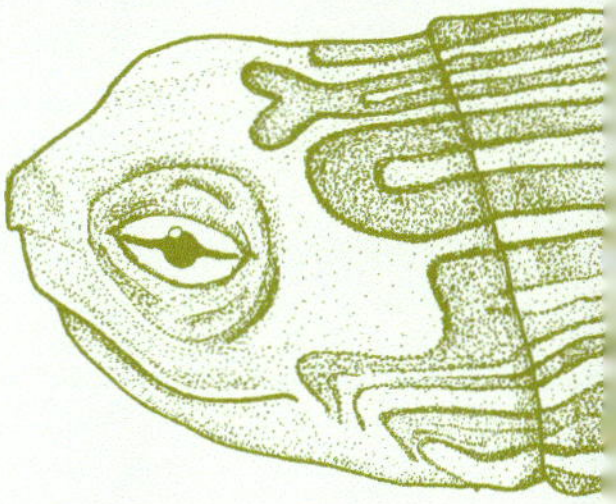

River, Pond, and Box Turtles

Family Emydidae

This family contains eleven genera and fifty-one species, many of which occur in North America. The family reaches its peak species richness in Alabama where seven genera and fifteen species are known to occur. These are mid-sized turtles that have hard shells and typically inhabit lakes, ponds, rivers, and sloughs. Most have colorful shells and bright striped, spotted, or blotched markings on the head and neck. Some genera, like *Graptemys,* have shells that are designed for swimming in swift waters, while others, like *Chrysemys,* have shells that are designed for swimming in stagnant water. Still others, such as *Malaclemys,* have invaded near-shore marine waters. Turtles in this family are sister to the families Geoemydidae + Testudinidae (Spinks et al. 2009). Based on a fossil specimen from the Pachuta Marl Formation of Choctaw County, the family has occupied Alabama for at least the last 35 million years.

Key to the Genera of Emydidae of Alabama

1a Plastron hinged.

Genus ***Terrapene*—American Box Turtles . . . . page 146.**

1b Plastron rigid. Go to 2.

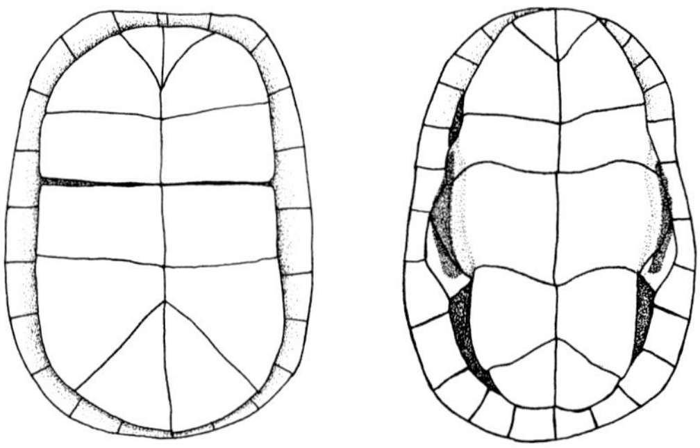

From left to right:

Hinged plastron of Box Turtle

Rigid plastron of Chicken Turtle

2a Head and neck with small gray spots; restricted to habitats associated with salt and brackish water.

Genus ***Malaclemys*—Diamond-backed Terrapins . . . . page 155.**

2b Head and neck with longitudinal light stripes except in some old melanistic individuals; found typically in freshwater habitats. Go to 3.

3a Dark bar or pair of large spots on bridge; carapace noticeably longer than wide and narrower at anterior than posterior end; neck long, when extended, distance from shoulder to snout tip approximately equals length of plastron.

Genus ***Deirochelys*—Chicken Turtles . . . . page 160.**

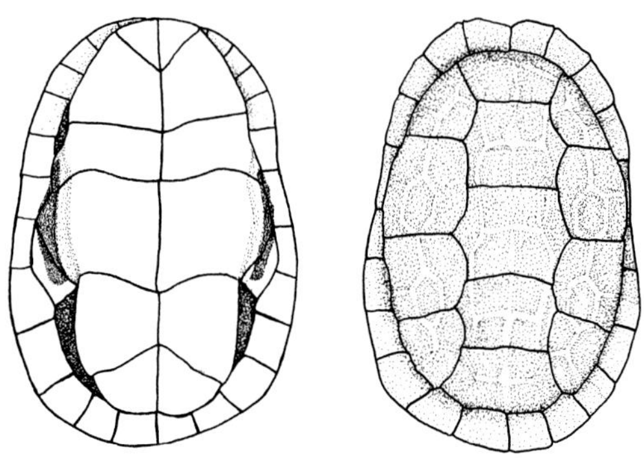

From left to right:

Plastron of Chicken Turtle with dark bar on bridge

Oblong carapace of Chicken Turtle

3b Bridge uniform yellow, lacking dark bar or spots; carapace round or wider at anterior end than posterior end. Go to **4**.

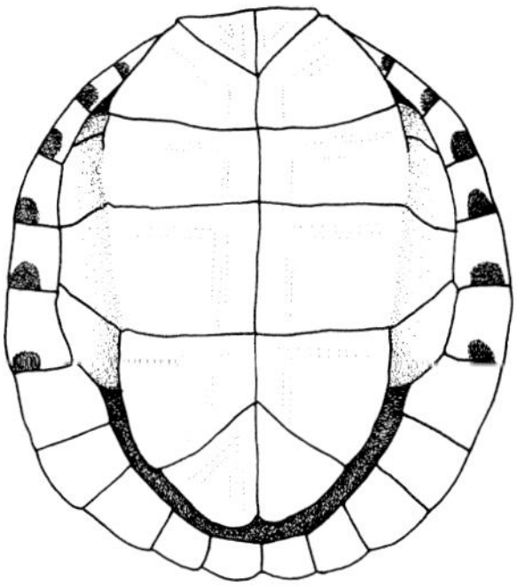

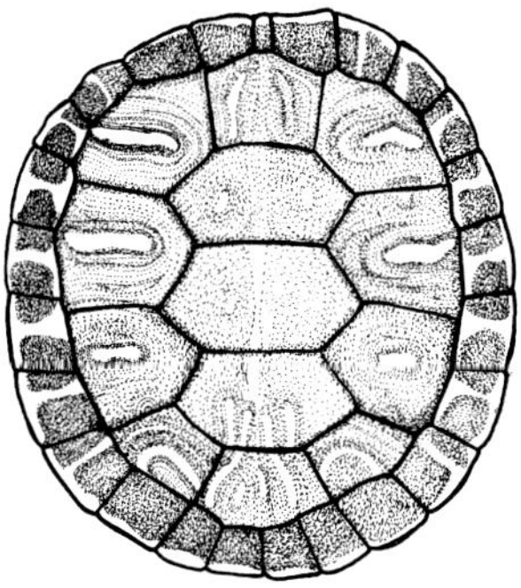

From left to right:

Plastron of Florida Cooter with uniform bridge

Round carapace of Florida Cooter

4a Carapace flat, smooth, and without a mid-dorsal ridge; pleurals olive to black, plain or light-rimmed.

Genus ***Chrysemys*—Painted Turtles . . . . page 164.**

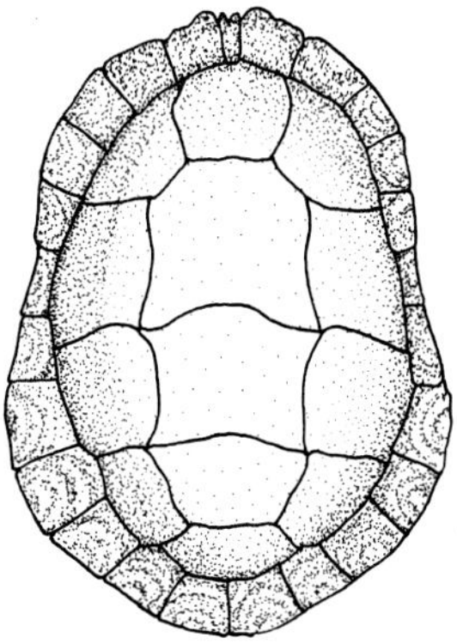

Flat, smooth carapace of Painted Turtle

4b Carapace with mid-dorsal ridge or with numerous shallow grooves, giving it a rough appearance. Go to **5**.

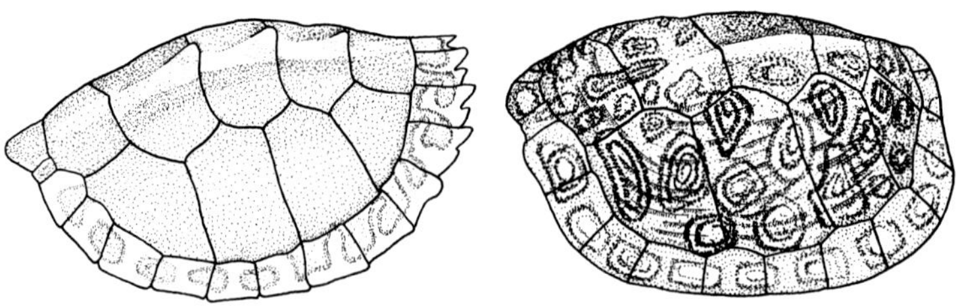

From left to right:

Carapace of Map Turtle with mid-dorsal ridge

Carapace of Cooter with numerous shallow grooves

5a Carapace wider anteriorly than posteriorly, with distinct keel mid-dorsally and serrate marginal scutes posteriorly.

Genus ***Graptemys*—Map Turtles . . . . page 172.**

5b Carapace round, without a mid-dorsal keel, lacking serrate edge to posterior marginals, and with a series of shallow grooves that give carapace a rough appearance. Go to **6**.

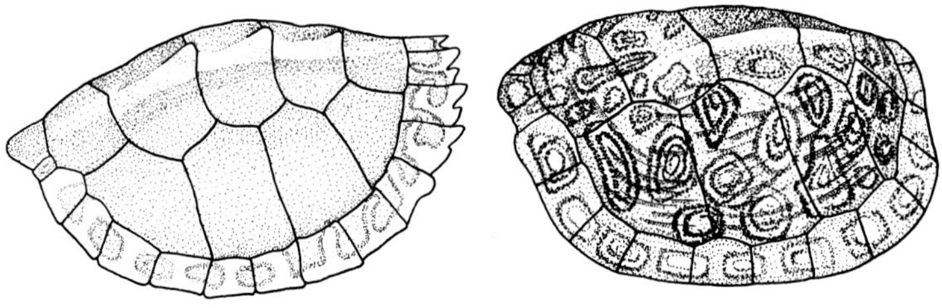

From left to right:

Carapace of Map Turtle with serrate marginal scutes

Rough carapace of Cooter without serrate edge

6a Plastron having two or more scutes marked in their centers with a dark smudge, blotch, or spot; plastron without dark markings following seams.

Genus ***Trachemys*—Sliders . . . . page 204.**

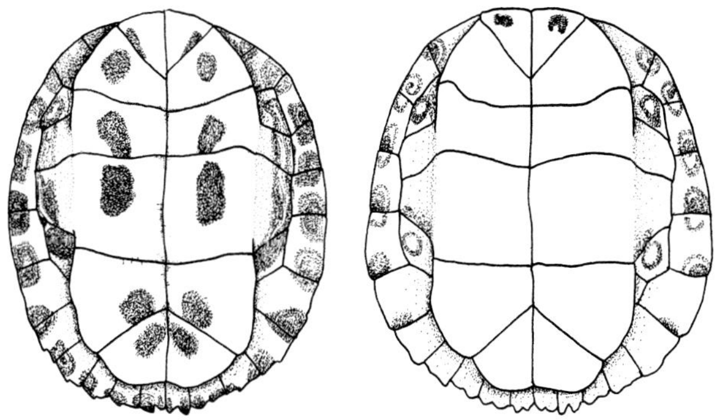

From left to right:

Plastron of Red-eared Slider with spots on all scutes

Plastron of Yellow-bellied Slider with with spots on anterior scutes

6b Plastron plain or, if marked, the markings following seams.

Genus ***Pseudemys*—Cooters . . . . page 213.**

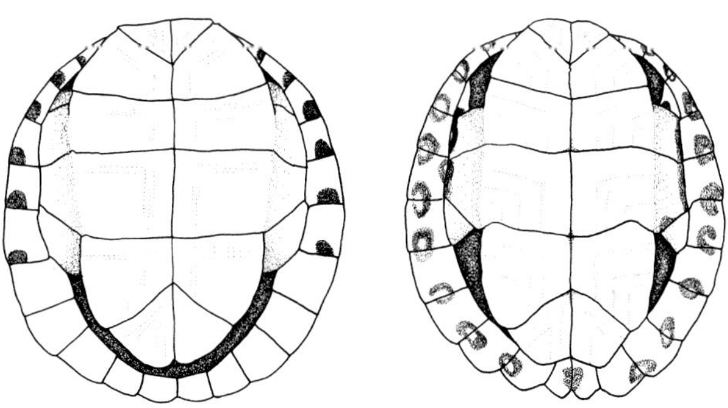

From left to right:

Plain plastron of Florida Cooter

Plain plastron of River Cooter

American Box Turtles

Genus *Terrapene* (Merrem, 1820)

Members of this genus are the basal lineage of the emydines, a radiation of semiterrestrial turtles within the family Emydidae (Stephens and Wiens 2003). The emydines are thought to have diversified in North America about 29 million years ago, sending descendants across the Bering land bridge to generate the ancestor to European and Asian lineages about 17 million years ago, and yielding the ancestral *Terrapene* about 12 million years ago (Spinks and Shaffer 2009). Known as Box Turtles, members of the genus are easily distinguished from all other Alabama turtles by the nature of the plastron, which is large, hinged, bilobed, and capable of completely sealing the turtle inside its shell.

Traditionally, *Terrapene* has been considered to include four species, two in Mexico and two in the United States. However, that taxonomy changed with the recent morphological and molecular work of Butler et al. (2011) on members of *T. carolina*. Because these authors found no structure to microsatellites of specimens of *T. c. major* and found no difference in morphological features of that taxon and the Pleistocene taxon *T. c. putnami*, they synonymized the two, with *T. c. major* having priority and being restricted to the fossil group rather than denoting a living taxon. Additionally, these authors elevated *T. c. baurii* to species status because it was shown to be a monophyletic lineage that is the sister taxon to *T. ornata*. Finally, Butler et al. (2011) documented that *T. c. triunguis* is a monophyletic lineage that is the sister taxon to *T. baurii* + *T. ornata*. Although they did not elevate *T. c. triunguis* to species status, we consider it to be a distinct species, leaving *Terrapene* as comprising six living species, two of which occur in Alabama. Mount (1975) considered much of the central portion of Alabama to show influence of a third species, *T. triunguis*. However, despite the presence of several Alabama specimens, Butler et al. (2011) found no evidence of the *T. triunguis* genome in the state and no Alabama specimen showed morphological features diagnostic of that taxon. This is surprising given that *T. triunguis* is documented to occur in Mississippi and Georgia (Butler et al. 2011). Based on these findings, we exclude *T. triunguis* from Alabama's turtle fauna.

Key to the Species of *Terrapene* of Alabama

1a Hind feet with four toes; carapace typically with conspicuous light markings, except in very old individuals, but these not forming two to five distinct radiating rays.

***Terrapene carolina*—Eastern Box Turtle . . . . page 148.**

1b Hind feet with three toes; carapace with light markings that form a series of two to five radiating rays.

***Terrapene bauri*—Florida Box Turtle . . . . page 152.**

Eastern Box Turtle

Terrapene carolina (Linnaeus, 1758)

Description This Eastern Box Turtle is a medium-sized terrestrial turtle that attains a maximum carapace length of around 8.5 inches (210 mm) in the extreme southern part of Alabama but that attains a maximum of 6.5 inches (165 mm) throughout the rest of the state. The shell lacks a bridge and is high-domed and rounded. The third vertebral scute is not elevated, so there is no hump shape to the carapace for specimens throughout most of the state. In the extreme southern part of Alabama the fifth vertebral scute may be convex, creating a hump shape to the carapace, and these specimens may also possess posterior marginals with a strongly flared, gutter-like conformation. The plastron is large, hinged, and capable of completely sealing the shell. The hind lobe of the plastron of adult males has a deep concavity (this region is flat in females). There are four toes on the hind feet, a feature that distinguishes this species from *T. bauri*. The carapace, head, and forelimbs are brownish to black with variable yellow or orange spots or slash markings; these can be bright in color, as though painted on the soft parts of the body. In areas where fire is used as a management tool, some individuals may be uniform olive-brown due to scarring

Left: Hatchling Eastern Box Turtle, Elmore County, AL

Opposite: Adult Eastern Box Turtle, Elmore County, AL

from being burned. The iris of the eye of males is usually red, while that of females is usually brown. The plastron of young Eastern Box Turtles consists of dark-centered markings that become highly variable with age. The carapace of the young are dark with a small yellow spot on each vertebral and pleural.

Alabama Distribution Analysis of morphology of specimens from Alabama indicates that this species is found throughout the state (Butler et al. 2011).

Habits Eastern Box Turtles are well known to Alabamians, who usually refer to them as "terrapins." They show little habitat specificity but are generally most abundant in moist, forested areas. During hot, dry weather they tend to move to floodplains, stream banks, and springheads and become somewhat sedentary. A light rain during a droughty period will often bring about a burst of Eastern Box Turtle activity, and at such times the turtles commonly are seen on roads.

The food requirements of Eastern Box Turtles, like their other ecological requirements, are seldom restrictive; berries, fungi, earthworms, insects, and even carrion are included in the diet.

During mating, the male mounts the female from the rear and

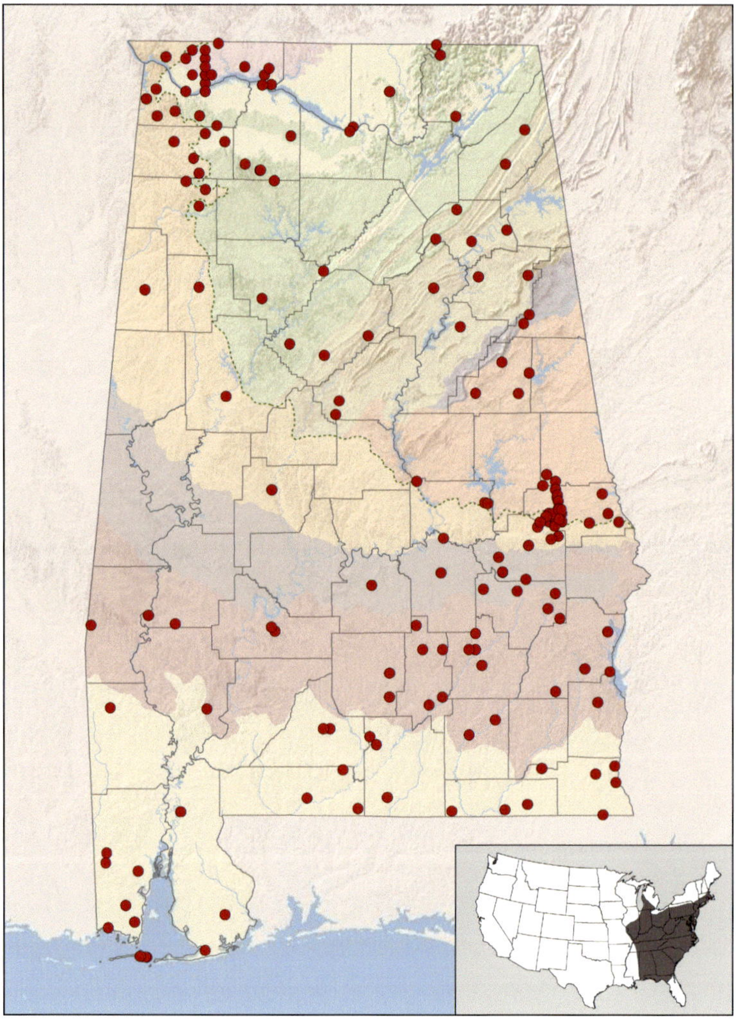

Alabama and continental US distribution of the Eastern Box Turtle

holds on with the hind feet. The concavity in the male's plastron facilitates assumption of the mating position. The female later deposits from two to seven elongate white eggs in a cavity she constructs in the soil. Apparently two or three three-egg clutches are laid per season. The sex of offspring from these eggs is determined by the temperature of the nest, with females produced from warmer nests and males from cooler nests (Janzen and Paukstis 1991). Curiously, baby Eastern Box Turtles are not often encountered, but when they are they tend to be under logs in moist habitats.

Conservation and Management Eastern Box Turtles are sold in the pet trade. Additionally, individuals in Alabama periodically contract a bacterial infection that causes severe distention of the ear region. An apparent decline in abundance associated with this health problem and collection in past years led to the provision in Alabama state law that prohibits possession of more than one Eastern Box Turtle per day for personal use. All Eastern Box Turtles receive some protection internationally through being listed as Lower Risk/Near Threatened by the IUCN. This species does well in regions where extensive woodlots are maintained. These areas can be urban, but road mortality can make it difficult to maintain populations in the long term. In forested areas that are fire-maintained, Eastern Box Turtles tend not to escape the flames. They may survive low-intensity fires, but often have extensive shell damage from the effects of these fires. A recently documented ranavirus (Johnson et al. 2008b) causes rapid death and should be considered a possible emerging threat to the species.

Taxonomy This species is found throughout the state. Based on the morphological and molecular work of Butler et al. (2011) we consider it to have no living subspecific variation. However, in the Coastal Plain of Alabama, *T. carolina* can be notably large, often with exceedingly flared posterior marginals. We follow Butler et al. (2011) in considering these to be specimens retaining alleles from the extinct *T. c. major.* Previous authors have placed this species in the genera *Cistudo, Emys, Monoclida, Terapene,* and *Testudo.*

Florida Box Turtle

Terrapene bauri (Taylor, 1894)

Description Florida Box Turtles are medium-sized terrestrial turtles that attain maximum carapace lengths of around 6.5 inches (165 mm). The shell lacks a bridge and is high domed and rounded. The fifth vertebral scute is convex, creating a hump shape at the posterior half of the carapace. The plastron is large, hinged, and capable of completely sealing the shell. The hind lobe of the plastron of adult males has a deep concavity (this region is flat in females). There are three toes on the hind feet, a feature that distinguishes this species from *Terrapene carolina*. The head and forelimbs are brownish to black with variable yellow or orange spots or slash markings. The carapace is dark brown with two to five bright yellow slash markings radiating from the center of each scute. These radiating lines differ from the haphazard yellow spotting of *T. carolina*. The iris of the eye of males is usually red, while that of females is usually brown. The plastron of young Florida Box Turtles consists of dark-centered markings that become highly variable with age. The carapace of young individuals is dark with a small yellow spot on each vertebral and pleural scute.

Alabama Distribution This species is known to occur in Alabama based on two specimens from Baldwin County and one from Houston

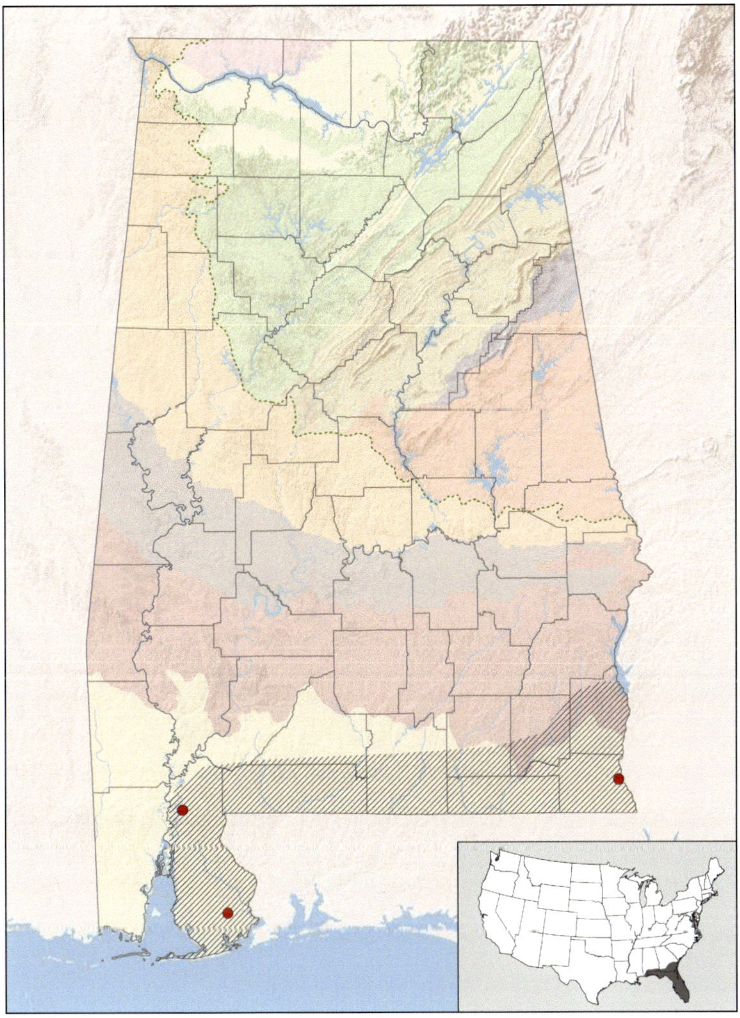

Left: Alabama and continental US distribution of the Florida Box Turtle

Opposite: Adult Florida Box Turtle, Jefferson County, FL

County. Butler et al. (2011) document specimens with the *T. bauri* genome from areas near the Escambia-Covington County line, suggesting that this species may be present along the southern tier of counties in Alabama.

Habits Florida Box Turtles are generally most abundant in moist, forested areas. During hot, dry weather they tend to move to floodplains, stream banks, and springheads and become somewhat sedentary. A light rain during a droughty period will often bring about a burst of

Florida Box Turtle activity, and at such times the turtles commonly are seen on roads.

Food items of Florida Box Turtles, like their other ecological requirements, are seldom restrictive; berries, fungi, earthworms, insects, and even carrion are included in the diet. This species is known to be a seed dispersal agent for some plants (Liu et al. 2004).

During mating, the male mounts the female from the rear and holds on with the hind feet. The concavity in the male's plastron facilitates assumption of the mating position. The female later deposits two to seven elongate white eggs in a cavity she constructs in the soil. Apparently, two or three three-egg clutches are laid per season. The sex of offspring from these eggs likely is determined by the temperature of the nest, with females produced from warm nests.

Conservation and Management Because this species used to be classified as a subspecies of *T. carolina*, it should be considered protected by Alabama state law; individuals may possess one individual per day for personal use. All Florida Box Turtles receive some protection internationally through their being listed as Lower Risk/Near Threatened by the IUCN. This species does well in regions where extensive woodlots are maintained. These areas can be urban, but road mortality can make it difficult to maintain populations in the long term. In areas that are fire-maintained, Florida Box Turtles survive better when burns occur during winter when the turtles are inactive (Platt et al. 2010). They may survive low-intensity fires, but often have extensive shell damage from the effects of these fires. A recently documented ranavirus (Johnson et al. 2008b) causes rapid death and should be considered a possible emerging threat to the species.

Diamond-backed Terrapins

Genus *Malaclemys* (Gray, 1844)

This genus consists of a single polytypic species whose range extends along the coast from New England to Mexico. Phylogenetically, this genus is most closely related to *Graptemys*, indicating an ancient split between this inhabitant of coastal estuarine habitats and the map turtle lineage occupying rivers that feed into these coastal areas (Stephen and Wiens 2003). Molecular evidence suggests that the ancestor of *Graptemys* and *Malaclemys* may have diverged from other emydids as long as 15 million years ago (Near et al. 2005). *Malaclemys* differs from its sister genus in retaining carapacial scutes so that each scute has a distinct pyramidal shape not seen in *Graptemys*, which sheds scutes during growth.

Taxonomy Seven subspecies are recognized, one of which occurs along the coast of Alabama. These subspecies show a major divergence between those that occur along the Atlantic Coastal Plain and those that occur along the Gulf Coastal Plain (Lamb and Avise 1992). However, specimens from Texas share mitochondrial genes with specimens from the Central Atlantic Coastal Plain, suggesting that movement of subspecies by humans created some populations (Hauswaldt and Glenn 2005). Previous authors have placed this species in the genera *Emys*, *Malacoclemmys*, and *Testudo*.

Mississippi Diamond-backed Terrapin

Malaclemys terrapin pileata (Wied-Neuwied, 1865)

Description Mississippi Diamond-backed Terrapins are medium-sized turtles of salt and estuarine marshes, attaining a maximum carapace length of about 9 inches (228 mm) in females and 6.5 inches (165 mm) in males. The shell of each individual is broadly oval, with sides that are nearly parallel. The carapace is variable in color but usually is dark gray to black without any bold markings in adults. The scutes of the carapace have deep furrows between them, creating conspicuous growth rings. All vertebrals except the last (fifth) have a medial swellings creating a weak keel; those on the third and fourth are prominent, rounded, and knoblike. The plastron is yellowish with a profusion of dark gray spots and blotches. These spots continue on the skin of the neck, chin, and cheeks, where they are distinctive against the light gray or greenish gray ground color. The upper lips and dorsal surface of the head are dark, as are the limbs. Juveniles are lighter in color with spotting that is more widely distributed. Adult females are larger in overall size than males, with the head of females being conspicuously enlarged.

Alabama Distribution Wetlands along the Gulf Coast of Baldwin and Mobile Counties.

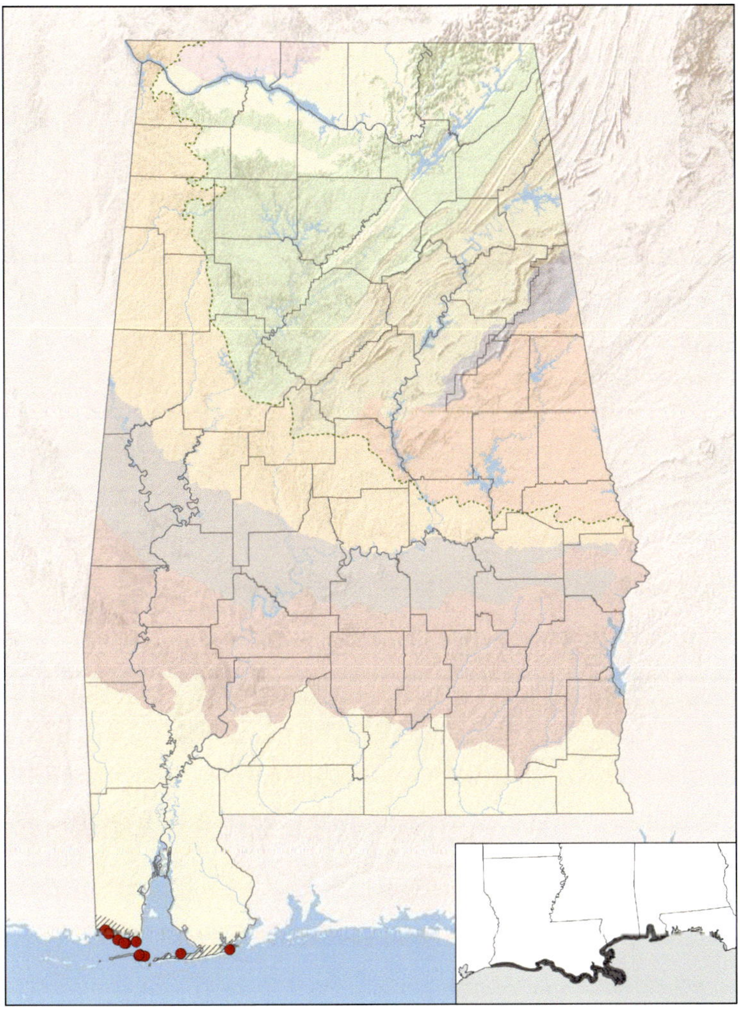

Left: Alabama and Gulf Coast distribution of the Mississippi Diamond-backed Terrapin

Opposite: Adult female Mississippi Diamond-backed Terrapin, Jackson County, MS

Habits This turtle is restricted to salt marsh and estuarine habitats. Adult females feed almost exclusively on mollusks, especially periwinkle snails, and crustaceans, especially blue crabs; these are crushed between the expanded alveolar surfaces of the jaws. Clutch size differs among years for Alabama terrapins, varying in mean clutch size from five to eight eggs. Elsewhere within the range of this species, eggs are laid in flask-shaped nests dug on high ground. Between two and five nestings occur each year for reproductive females. The sex of offspring from these eggs is determined by the temperature of the nest, with females produced from warmer nests and males from cooler nests (Janzen and Paukstis 1991).

Conservation and Management The Mississippi Diamond-backed Terrapin is ranked Priority 1 (Highest Conservation Concern) by ADCNR (2012 modification of Mirarchi et al. 2004) and, therefore, is protected by the state of Alabama. Diamond-backed Terrapin flesh is considered highly flavorful and prized for use in soups and stews. This human use was strong enough that an entire fishery developed along the Atlantic and Gulf Coasts to provide terrapins for restaurants and canneries. One of the largest ranching operations in the 1880s was along Alabama's Gulf Coast just north of Dauphin Island, providing as many as 10,000 terrapins per year to New York importers. This level of harvest depleted the resource throughout the range of the entire species, leading to conservation legislation in most states to protect the species.

Given that terrapins have not been harvested in great numbers in Alabama for decades, it is surprising that this species has not yet recovered. Mount (1975) speculated that Hurricane Camille in 1969 removed the high dunes that served as a primary nesting site for the terrapins in the state. However, adult terrapins drown in crab traps, and this fishery is common along the coast of Alabama. Adult mortality thus continues to deplete the reproductive stock of this turtle. Finally, nest predators, especially raccoons and fish crows, are abundant in Alabama and consume a large proportion of nests in the state.

Recent surveys (Wibbels 2010) failed to find terrapins in the Bon Secour area of Baldwin County, but over a three-year period approximately 100 Mississippi Diamond-backed Terrapins were marked and released along the coast in Mobile County. Recent sightings have documented the species to be present at Barton Island, Point of Pines, Mon Luis, Airport Marsh (Dauphin Island), Jemison's Marsh, Cedar Point Marsh, and Little Dauphin Island (Wibbels 2010). Of these, the largest numbers, and most consistent sightings, have occurred at Cedar Point Marsh, where an estimated population of fifty adult females resides. At this location a region of shell hash beach is the most persistently used nest site in the state, with 100–150 nests being created per year. Airport Marsh also serves as an important nest site, with approximately forty nests per year. These two sites are the primary focus of conservation efforts in the state.

Two primary management activities are being tested for use in increasing Mississippi Diamond-backed Terrapin populations in Alabama. One is headstarting eggs gathered from Cedar Point Marsh. In this program, eggs are incubated to hatching at the University of Alabama at Birmingham, with the turtles raised to approximately one year of age and then released at Cedar Point Marsh. A pilot project in 2010 generated forty-eight individuals that have been released as a first cohort, with a target of releasing approximately 200 individuals by the end of an initial three-year project. The second management activity that is being tested is the use of turtle excluder devices on crab traps along the Gulf Coast of Alabama. These devices reduce capture rates of terrapins by 80 percent while not affecting the rate of crab capture; they are required in some states within the range of diamond-backed terrapins. Their required use within Alabama's waterways appears vital to restoring healthy populations of Mississippi Diamond-backed Terrapins. Alternatively, banning the use of crab traps near the Cedar Point Marsh and Airport Marsh sites should benefit populations at those sites; creation of such sanctuaries is listed as a conservation goal in the SWAP (Alabama Department of Conservation and Natural Resources, Division of Wildlife and Freshwater Fisheries 2005). Because nest predation appears to be so high, control of raccoon populations near Cedar Point Marsh and Airport Marsh also appears to be a beneficial management activity for allowing recovery of Mississippi Diamond-backed Terrapins in the state.

Chicken Turtles

Genus *Deirochelys* (Agassiz, 1857)

The Chicken Turtle derives its common name from the fact that its meat is tasty to humans. The single species placed in this genus is limited almost entirely to the southeastern Coastal Plain. It is characterized by an extremely long neck that is used to capture crayfish. *Deirochelys* is the sister taxon to the ancestor of *Graptemys* + *Malaclemys* + *Trachemys* + *Pseudemys* (Spinks et al. 2009). Along with *Chrysemys,* these genera appear to have occupied river and pond habitats of North America for as long as 30 to 40 million years (Spinks and Shaffer 2009). Members of the genus are smaller in northern populations than in southern populations (Ashton and Feldman 2003).

Eastern Chicken Turtle

Deirochelys reticularia reticularia (Latreille, 1801)

Description Eastern Chicken Turtles are medium in size, attaining a maximum carapace length of about 10 inches (255 mm). The neck is noticeably long when extended, sometimes approaching that of the carapace in its length. The carapace is dark, oval to teardrop in shape, marked with a typically inconspicuous reticulum of greenish or yellowish lines, and often rimmed with yellow. The plastron is rigid, yellow, typically unmarked or, when marked, has a dendritic pattern following the seams. The bridge usually has an elongate black blotch that may be broken into two or three spots. The marginals near the bridge often have dark spots on their undersurfaces. Ground color of the soft parts is dark brown or black and the forelimbs have a broad yellow stripe on the lower edge of the anterior surface. The posterior surface of each thigh is marked with a series of vertical yellow stripes, separated by dark greenish brown. Chicken Turtles are most similar to members of the genus *Pseudemys*, but those turtles lack the elongate neck and black blotches on the bridge of *Deirochelys*.

Alabama Distribution Eastern Chicken Turtles can be found from the Coastal Plain, where they are common, to the Ridge and Valley region, where they are encountered infrequently.

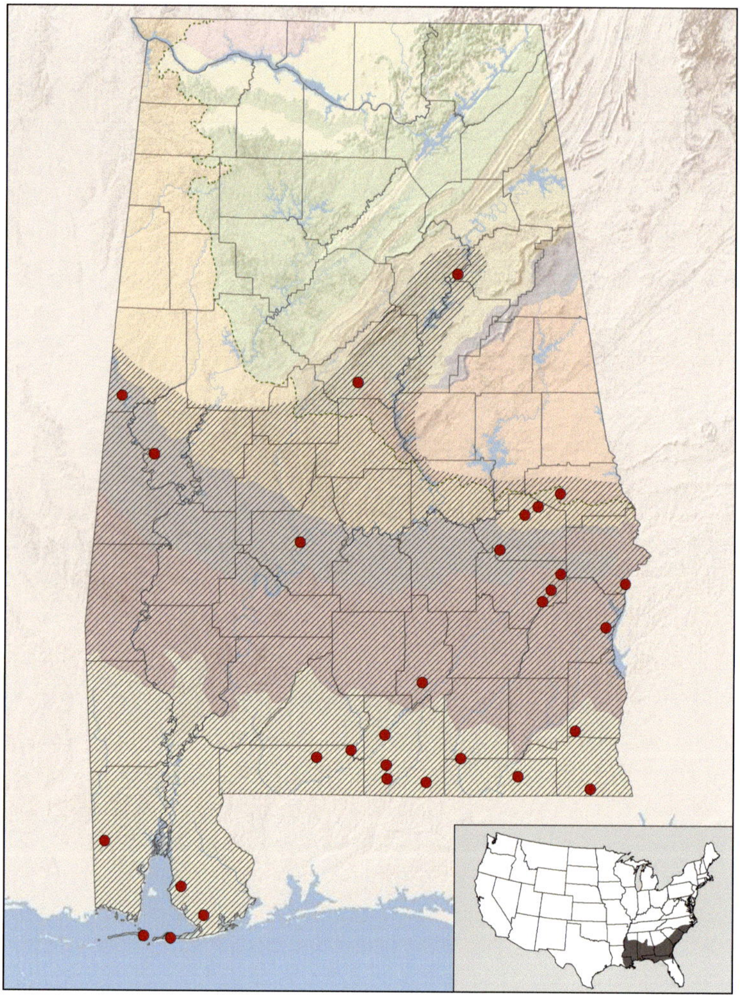

Right: Alabama and continental US distribution of the Eastern Chicken Turtle

Previous: Adult Eastern Chicken Turtle, Macon County, AL

Habits Shallow, weedy ponds, swamps, and borrow pits with standing water are the usual habitats of the Eastern Chicken Turtle. It is not strongly aquatic and often moves about on land, especially during spring, and prefers smaller bodies of water that do not have large predatory fishes or alligators. Such wetlands frequently dry, and adult turtles in such situations move to upland sites where they aestivate under leaf litter (Buhlmann 1995). Male courtship involves swimming in front of a female with arms extended and scratching the head and neck of the female with the male's claws. If a female is receptive, then mating occurs. Females then deposit eggs in a nest either in August to September or during February to May (Congdon et al. 1983). This process involves either generating a clutch of eggs in late summer, retaining that clutch through winter and laying it early the next year, or generating a clutch in late summer and depositing it that fall (Buhlmann et al. 1995). Clutch size is five to nineteen eggs (Buhlmann et al. 2008), and these go through a period of diapause in which growth is suspended until they pass through a cool period that triggers resumption of development.

Eastern Chicken Turtles are suction feeders, consuming mostly aquatic invertebrates. Individual turtles wait for prey to approach closely, and then shoot the head and neck away from the body while opening the mouth and neck. This process sucks in water and prey. The water is expelled from the mouth and the prey is swallowed whole. Juveniles and males feed mainly on dragonfly larvae while females consume crayfish.

Conservation and Management To protect this species from unsustainable harvest, no more than two individuals of Eastern Chicken Turtles can be possessed, and these must be for personal use and not for sale. This species occupies the kinds of wetlands that can be easily altered by human activities. Because these wetlands tend to be small, they also tend to dry frequently. When this happens, Eastern Chicken Turtles either move to alternate wetlands or into upland habitats where they bury themselves and wait for rains to refill the wetland. For this reason, typical management zones around wetlands will be insufficient to retain this species if land management around wetland sites involves frequent disturbance by heavy equipment. Prescribed burning around wetlands, while generally beneficial, can kill Eastern Chicken Turtles if conducted when the animals are dormant under leaf litter during the winter or when ponds are completely dry in summer. Because of this and its spotty distribution within the state, the species is listed as vulnerable on the Alabama Natural Heritage Program tracking list. In addition, wildlife fencing and designed underpasses in areas of heavy traffic may be necessary to maintain the species in urban settings because Eastern Chicken Turtles, especially the females, cross roads frequently (Steen et al. 2006).

Taxonomy Three subspecies of this southeastern turtle are recognized, one of which occurs in Alabama. Previous publications place this species in the genera *Chrysemys*, *Emys*, and *Testudo*.

Painted Turtles

Genus *Chrysemys* (Gray, 1844)

This genus contains the Painted Turtles, chelonians with low, flat, shiny shells and with combined ranges covering most of the United States. There are two extant species, both of which are found in Alabama. Painted Turtles are the sister genus to the common ancestor of river and pond turtles of North America (*Pseudemys, Trachemys, Graptemys, Malaclemys,* and *Deirochelys*; Spinks et al. 2009). Members of the genus are larger for northern populations, subspecies, and species than in southern forms (Ashton and Feldman 2003). These turtles occupy sluggish or stagnant waters that they frequently share with the genus *Trachemys*. Painted Turtles frequently bask in such habitats where their extremely shiny shell makes them distinguishable from *Trachemys,* which have numerous shallow grooves on their carapace that give them a dull appearance when basking.

Key to the Species of *Chrysemys* of Alabama

1a Carapace with a prominent yellow, orange, or red mid-dorsal stripe.
   ***Chrysemys dorsalis*—Southern Painted Turtle . . . . page 165.**

1b Carapace without a prominent mid-dorsal light stripe.
   ***Chrysemys picta picta*—Eastern Painted Turtle . . . . page 168.**

Southern Painted Turtle

Chrysemys dorsalis (Agassiz, 1857)

Description A mid-sized turtle with a maximum carapace length of about 6 inches (150 mm), this species is characterized by the presence of a conspicuous red, orange, or yellow median stripe on the carapace. The vertebral and pleural scutes of the carapace have sutures that do not align with each other, instead alternating from anterior to posterior. The head is dark with bright yellow markings that form longitudinal stripes. A yellow stripe or elongate mark is found on the head behind each eye. Lighter markings on the legs and other soft parts often are red. The Eastern Painted Turtle differs from this species by lacking a light mid-dorsal stripe and possessing distinct round spots behind each eye.

Alabama Distribution Southern Painted Turtles are found throughout the Tennessee River system but show evidence of past hybridization with *Chrysemys picta picta* in Jackson County. They also are found in that portion of the Mobile Bay drainage located below the Fall Line.

Habits Southern Painted Turtles are locally abundant in Alabama in lakes, ponds, sloughs, and oxbows that have mud or silt bottoms and abundant aquatic vegetation. They also occur in the main channels of

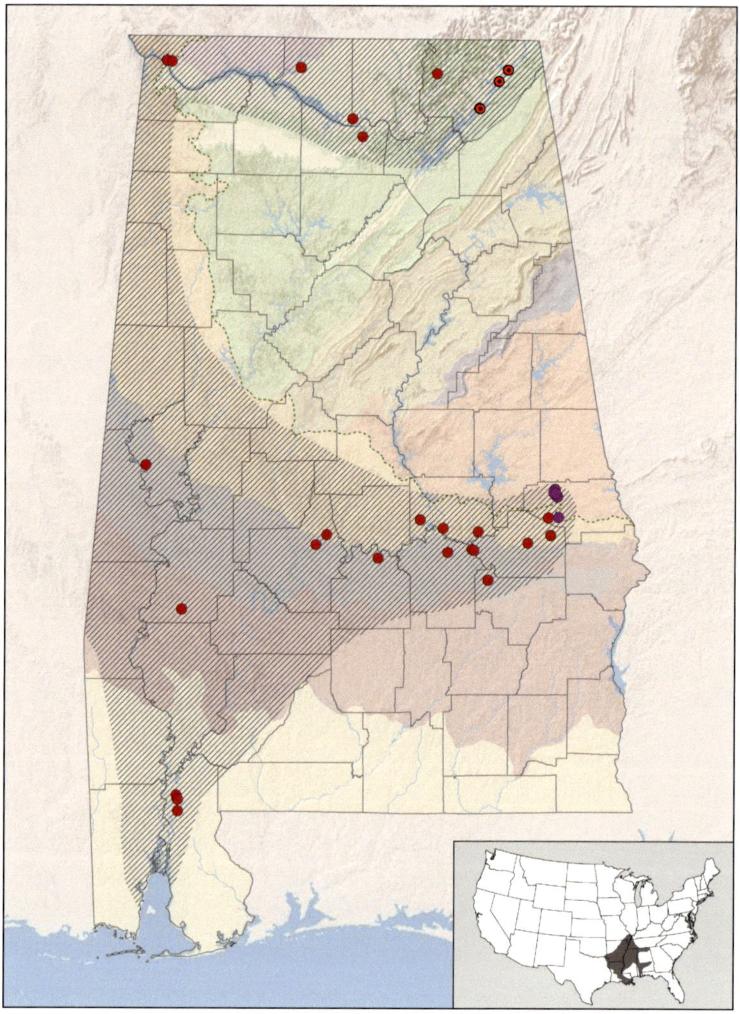

Right: Alabama and continental US distribution of the Southern Painted Turtle; purple dots indicate hybrids with Eastern Painted Turtle

Previous: Adult Southern Painted Turtle, Madison County, AL

some of our large sluggish streams, but they are seldom abundant in those habitats. During winter, these turtles burrow in mud in shallow water during cold weather, but emerge again in warmer weather. During the active season (May-October) the basking habit is strongly developed in Southern Painted Turtles, and they are often found wandering on land. Food habits are generalized, with arthropods and mollusks predominating, but also including substantial amounts of aquatic vegetation, especially duckweed (*Lemna*). Males court females in spring, when mating occurs. Females then usually dig their nest cavities sometime between mid-May and the end of June. Nest sites are open areas with loose soil, where females select warm, well-drained locations and nest temperature determines offspring sex (Schwarzkopf and Brooks 1987). Data on clutch size from Alabama are lacking, but elsewhere clutches for this species average four eggs and females typically produce two clutches per year.

Conservation and Management This is a common species that survives well in urban settings. Therefore, no special conservation efforts are required for Southern Painted Turtles. However, to protect this species from unsustainable harvest, no more than two individuals of Southern Painted Turtles can be possessed and these must be for personal use and not for sale. Management practices that increase the presences of basking sites (logs or similar isolated solid objects jutting above the water line) and streamside management zones that protect riparian forests are likely to improve the habitat for this species.

Taxonomy We follow Starkey et al. (2003) in elevating this lineage to species status. Until that publication *C. dorsalis* was recognized as a subspecies of *C. picta*. Previous publications have placed this species in the genera *Emys*, *Hydrochelys*, *Testudo*, and *Trachemys*.

Eastern Painted Turtle

Chrysemys picta picta (Schneider, 1783)

Description This is a medium-sized aquatic turtle attaining a maximum carapace length of about 7 inches (180 mm). The carapace is smooth, dark, and somewhat depressed, with the seams of the second and third vertebral scutes aligning so that they are nearly in the same transverse plane as those of adjacent pleurals. These seams frequently are light yellow in color, creating thin light bands across the carapace. The plastron is yellowish and usually is unmarked. Conspicuous red marks are found on the marginals. A pair of yellow spots is found on the head behind the eyes, and another pair occurs medially and posteriorly to these. Lighter markings on the legs and other soft parts are red. This species differs from *C. dorsalis*, which has a light mid-dorsal stripe on the carapace and has yellow stripes or elongate markings behind the eyes.

Alabama Distribution Eastern Painted Turtles are found in the Chattahoochee, Tallapoosa, and Coosa Rivers above the Fall Line. No Eastern Painted Turtles are found in the Chattahoochee drainage below the Fall Line and *C. dorsalis* replaces *C. p. picta* below the Fall Line in the drainages of the Alabama River. Populations of apparent *C. dorsalis* x *C. p. picta* hybrids are found in North Auburn Fisheries ponds associated

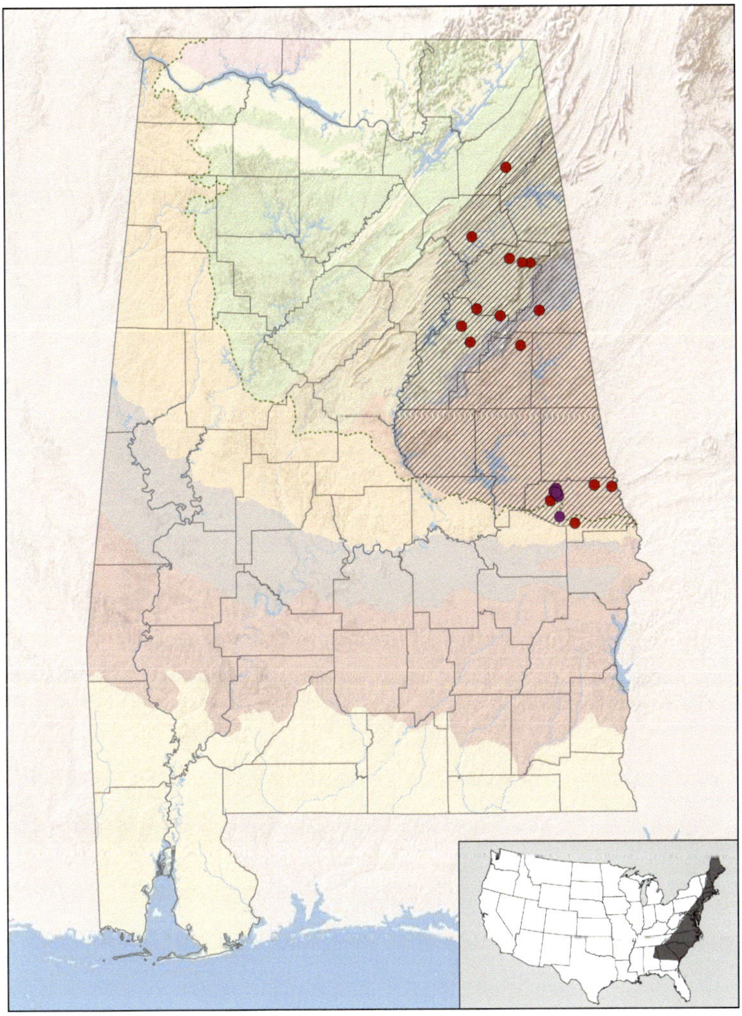

Left: Alabama and continental US distribution of the Eastern Painted Turtle; purple dots indicate hybrids with the Southern Painted Turtle

Opposite: Adult Eastern Painted Turtle, Lee County, AL

with tributaries of Saugahatchee Creek, just north of the Fall Line in Lee County. We suspect, however, these have been influenced by humans moving *C. dorsalis* into these ponds and that similar hybrid populations occur in other streams associated with the Fall Line.

Habits Eastern Painted Turtles inhabit lakes, ponds, sloughs, and oxbows having mud or silt bottoms and abundant aquatic vegetation. During winter these turtles burrow into the muddy bottom of shallow wetlands, where they can survive for up to 150 days without feeding or breathing, a duration that is superior to that known for *C. dorsalis* (Ultsch et al. 1999). This subspecies can be locally abundant in these habitats and are less abundant in the main channels of some of our large sluggish streams. The basking habit is strongly developed in this subspecies, and it is often found wandering on land. Food habits are generalized, with arthropods and mollusks predominating, but likely with substantial amounts of aquatic vegetation being consumed.

Males court females in spring, when mating occurs. Females then usually dig their nest cavities between mid-May and the end of June, when they lay four to eleven eggs. Nests usually are placed on sandy beaches, and females are capable of producing two clutches per year. The sex of offspring from these eggs is determined by the temperature of the nest, with females produced from warmer nests and males from cooler nests (Janzen and Paukstis 1991). Early nests typically hatch and emerge in fall, but late nests may overwinter before emerging in spring.

Conservation and Management This species has a fairly restricted geographic range within Alabama, but is widespread elsewhere. It is common in urban settings and therefore requires no special conservation efforts. However, to protect this species from unsustainable harvest, no more than two individuals of Eastern Painted Turtles can be possessed, and these must be for personal use and not for sale. Populations in areas with high road densities become strongly male-biased because of increased female mortality associated with individuals killed by vehicles while searching for an appropriate nest site (Steen et al. 2006). Therefore, carefully designed wildlife fences and underpasses for this species are worth including in urban planning. Additionally, management practices that increase the presences of basking sites (logs or similar isolated solid objects jutting above water line) are likely to improve the habitat for this species, as will implementation of streamside management zones appropriate for wildlife.

Taxonomy We follow Starkey et al. (2003) in recognizing *C. dorsalis* as a separate species rather than as a subspecies of *C. picta* (Mount 1975). According to this taxonomy, three subspecies of *C. picta* remain, one of which occurs in Alabama. Mount (1975) reported *C. p. marginata* to be present in the state based on two specimens from extreme northeastern Alabama. However, the only specimens in the Auburn

University collections with the dark ventral figure expected of this subspecies are hatchlings from an area in Lee County (North Auburn Fisheries Ponds) where apparent *C. dorsalis* x *C. picta* hybrids are common. Additionally, a monophyletic lineage, based on the mitochondrial genome, which appears to be consistent with the putative range of *C. p. marginata,* occurs far north of the Alabama-Tennessee border (Starkey et al. 2003). Therefore, we interpret all Alabama specimens with a dark ventral figure to be hybrids between *C. dorsalis* and *C. picta picta* rather than documenting the influence of *C. p. marginata* in the state. Previous publications may place this species in the genera *Emys, Hydrochelys, Testudo,* and *Trachemys.*

Map Turtles

Genus *Graptemys* (Agassiz, 1857)

This genus contains the Map and Sawback turtles, so named because of intricate, map-like light markings on the carapaces of some species or the serrate mid-dorsal keel of the carapace of others. Thirteen species, all found in the eastern half of the United States and in a small adjacent portion of Canada, are placed in this genus. The sister taxon to *Graptemys* is *Malaclemys*, a lineage that has invaded near-shore marine environments and differs from *Graptemys* in possessing pyramidal plastral scutes; the common ancestor of these two genera may have diverged from other emydids 15 million years ago (Near et al. 2005). *Graptemys* displays unusual diversity in head morphology, especially of females, and exceptional sexual size dimorphism, with males being much smaller and retaining smaller head sizes than females (Lindeman 2008). Members of the genus are divisible into three phylogenetic groups: a narrow-headed group, containing *G. caglei, G. flavimaculata, G. nigrinoda, G. oculifera, G. ouachitensis, G. pseudogeographica,* and *G. versa*; a monotypic group containing *G. geographica*; and a broad-headed southern group containing *G. barbouri, G. ernsti, G. gibbonsi, G. pearlensis,* and *G. pulchra* (Stephens and Wiens 2003). All three phylogenetic groups are represented by the six species of *Graptemys* occurring in Alabama. Based on data for *G. pseudogeographica,* members of the genus can be larger in northern populations than in southern ones (Ashton and Feldman 2003).

Key to the Species of *Graptemys* of Alabama

1a Head with a round, isolated, white or yellow temporal spot.

Graptemys geographica—**Northern Map Turtle . . . . page 175.**

1b Head variously marked, but lacking a round, isolated, white or yellow temporal spot (except occasionally in *Graptemys ouachitensis*). Go to **2.**

2a Top and side of head and cheek with large yellow or greenish blotches; obscure in old females; upper jaw without a terminal notch. Go to **3.**

2b Top and side of head and cheek lacking large, light blotches or, if blotches present, upper jaw with a terminal notch. Go to **5.**

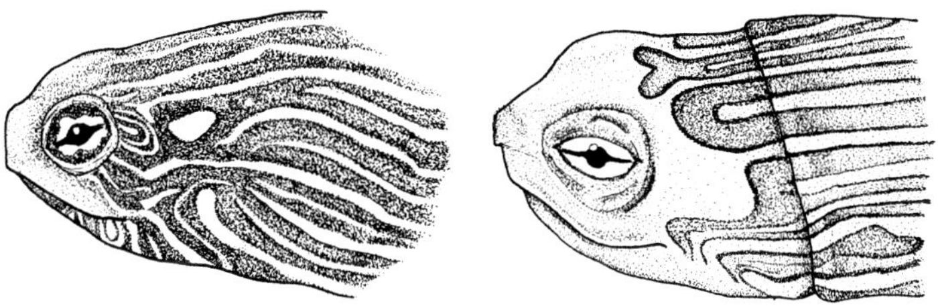

From left to right:

Head pattern of Northern Map Turtle with temporal spot

Head pattern of Alabama Map Turtle without temporal spot

3a Chin with a U-shaped light bar that does not connect to a median longitudinal chin stripe.

Graptemys barbouri—**Barbour's Map Turtle . . . . page 179.**

3b Chin with a V-shaped light bar that usually is connected to a median longitudinal light chin stripe. Go to **4.**

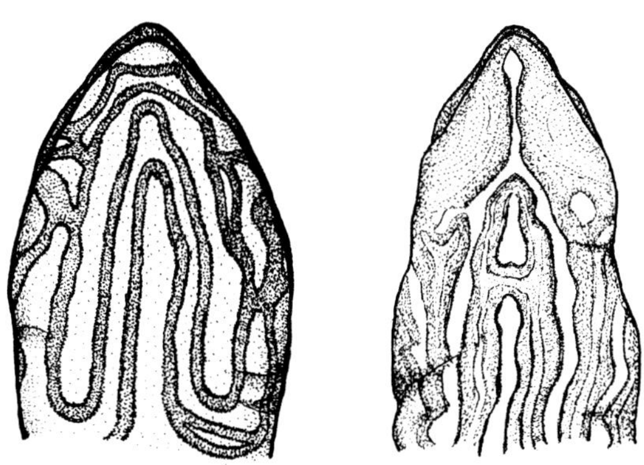

From left to right:

Chin pattern of Barbour's Map Turtle with U-shaped bar

Chin pattern of Escambia Map Turtle with V-shaped bar

4a Interorbital blotch in contact with postorbital blotches.

Graptemys pulchra—**Alabama Map Turtle . . . . page 183.**

4b Interorbital blotch separate from or narrowly connected to postorbital blotches.

Graptemys ernsti—**Escambia Map Turtle . . . . page 187.**

From left to right:

Head pattern of Alabama Map Turtle with connected orbital blotches

Head pattern of Escambia Map Turtle with disconnected orbital blotches

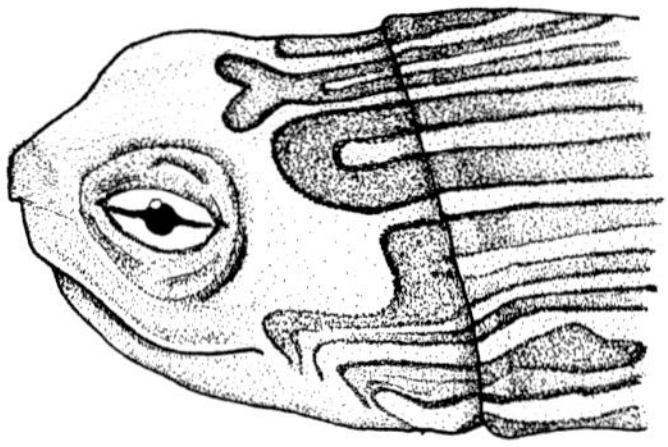

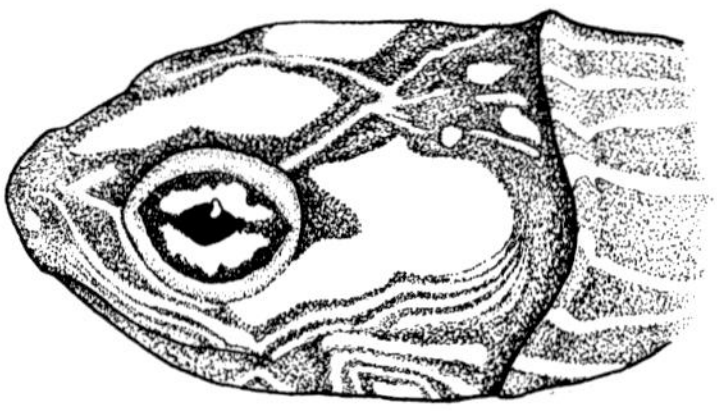

5a Chin with three conspicuous isolated light spots.

Graptemys ouachitensis—**Ouachita Map Turtle . . . . page 192.**

5b Chin variously marked, but without three conspicuous isolated light spots.

Graptemys nigrinoda spp.—**Black-knobbed Sawback Turtles . . . . page 195.**

Chin pattern of Ouachita Map Turtle with three chin spots

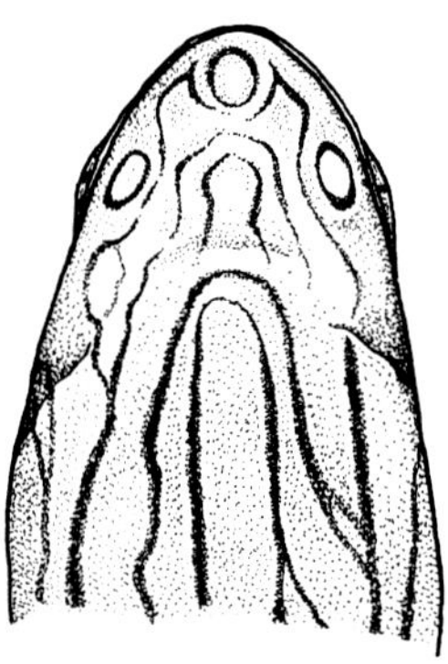

Northern Map Turtle

Graptemys geographica (Le Sueur, 1817)

Description This species is a large aquatic turtle attaining a maximum carapace length of about 10.5 inches (270 mm) in females and 6 inches (150 mm) in males. The carapace is relatively low and has a weak mid-dorsal keel along the vertebrals. Low spines or tubercles are found on the rear of some vertebrals in juveniles. In contrast with other members of the genus, the carapace is only weakly serrate along its posterior border. Green and olive brown are the dominant colors of the carapace, with a pattern of circular, semicircular, and reticulate yellow or orange markings creating a map-like appearance. The plastron is yellowish with dark lines along the seams and occasionally with other dark markings toward the center of the plastral scutes. The undersurfaces of the marginals have concentric circular or semicircular dark markings at the seams, and the bridge has dark, longitudinal lines. The head is greenish to olive with yellow stripes and a detached yellow spot behind the eye; these spots are unique among Alabama members of the genus *Graptemys*. In adult females, the head is considerably enlarged relative to males of similar age.

Alabama Distribution This species inhabits streams of the Tennessee River system and is found in the Black Warrior, Cahaba, and Coosa River systems above the Fall Line.

Above: Adult female Northern Map Turtle, Marion County, TN

Previous: Juvenile Northern Map Turtle, Winston County, AL

Habits This shy turtle is seen most frequently basking on logs and rocks in creeks and rivers. It is positively associated with stream stretches that are bordered by forested areas and avoids unforested areas; the species also overwinters in areas of streams bordered by forest (Ryan et al. 2008). Unlike other Alabama species of *Graptemys, G. geographica* will, on occasion, move well up into small brooks. The sex of offspring emerging from eggs is determined by the temperature of the nest, with females produced from warmer nests and males from cooler nests (Janzen and Paukstis 1991). Juveniles and adult males feed on a wide variety of small aquatic invertebrates. Mature females feed mostly on mussels and snails. *Graptemys geographica* is not particularly common anywhere in Alabama, and data on its reproduction in the state are lacking. Elsewhere it is known to lay ten to sixteen elliptical, leathery-shelled eggs during the morning hours in May and June.

Conservation and Management Northern Map Turtles, like all turtles in Alabama, may not be captured for sale. A person may possess a maximum of two individuals per day for personal use only. This

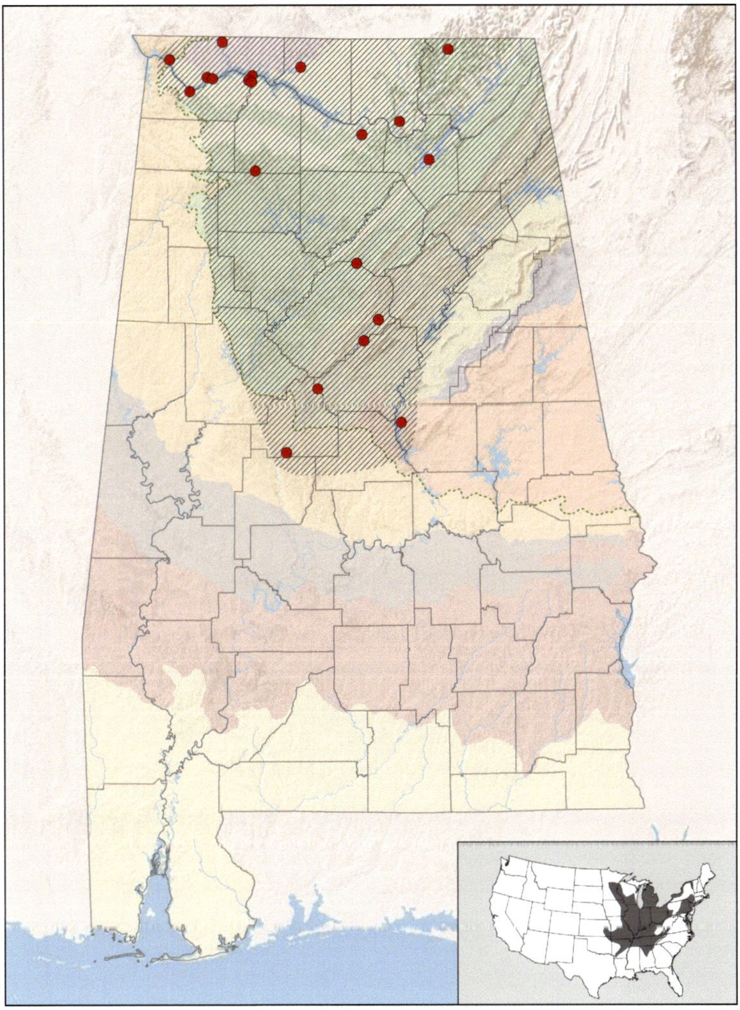

Alabama and continental US distribution of the Northern Map Turtle

species is widespread and common outside the state of Alabama, where the core of its geographic range occurs. Because population density tends to become reduced near the edge of a species' range, the low population densities of the species in Alabama are not necessarily due to human activities. However, the species is listed as vulnerable in the Alabama Natural Heritage Program tracking list. Maintenance of clear, rocky streams with protected streamside management zones and retention of snags in streams are likely to be sufficient to maintain populations of this species in the state. Any conservation activities that preserve current stocks of native mussels or restores past mollusk beds will enhance populations of *G. geographica*. Additionally, because members of this species, especially the females, cross roads frequently (Steen et al. 2006), wildlife fencing and designed underpasses in areas of heavy traffic may be necessary to maintain the species in urban settings.

Taxonomy No subspecific variants are recognized for this species. This species is the sister taxon to the common ancestor of the narrow-headed (*G. flavimaculata* + *G. oculifera* + *G. nigrinoda* + *G. pseudogeographica* + *G. ouachitensis* + *G. versa*) and broad-headed (*G. barbouri* + *G. ernsti* + *G. gibbonsi* + *G. pulchra* + *G. caglei*) map turtles (Stephens and Wiens 2003). Previous authors have placed this species in the genera *Emys*, *Malacoclemmys*, and *Testudo*.

Barbour's Map Turtle

Graptemys barbouri (Carr and Marchand, 1942)

Description Barbour's Map Turtles are large aquatic organisms that attain a maximum carapace length of about 13 inches (330 mm) in females and 5 inches (130 mm) in males. The carapace has a median keel accentuated by spines or prominent knobs on some of the vertebrals; these are most pronounced on the second and third vertebrals and become inconspicuous on large females. At its posterior end, the carapace is serrate. The ground color of the carapace is olive or greenish with the tips of the spines or knobs being black. The pleural scutes typically have yellowish circular or C-shaped markings and the upper surfaces of the marginals have narrow yellowish markings that may be square in shape. On their ventral surface, the marginals and bridge have dark circular markings near the posterior suture of each scute. Carapaces of adult females darken with age so that the markings become obscure. The plastron is uniform yellow, except for narrow dark lines following the seams. The head has a large greenish or yellow-green blotch behind each eye and a similarly colored blotch on the top of the head between the eyes. A wide, U-shaped yellow bar is found on the undersurface of the chin (to distinguish this species from *G. ernsti* and *G. pulchra*, which have a V-shaped light bar). Both the limbs and the tail are striped with yellow.

Right: Adult female Barbour's Map Turtle, Baker County, GA

Previous: Adult male Barbour's Map Turtle, Baker County, GA

Alabama Distribution The species is found in the Chattahoochee River, northward at least to Halawakee Creek in Lee County. It is also found in the Choctawhatchee and Pea Rivers. It has not been taken from Big Creek or Cowart's Creek in southern Houston County, which are headwaters of the Chipola River, a Florida river that contains *G. barbouri.*

Habits Barbour's Map Turtle is found exclusively in streams that contain substantial numbers of mollusks. It is fond of basking, utilizing for this purpose logs, snags, stumps, and the slanting bases of living trees that protrude from the water. When they bask, adult females are exceedingly wary and difficult to approach. However, they are easily caught underwater in clear streams by competent swimmers who follow individuals to their underwater retreats (submerged logs or rocks); under some conditions these turtles may be taken at night while they sleep near the surface. Adult females have home ranges that average about 4 acres (1.7 ha), and these home ranges include a deep pool and exposed and submerged logs as key resources (Sterrett 2009).

Barbour's Map Turtle is wholly carnivorous. Adult males and juveniles of both sexes are mostly insectivorous, while adult females, with their enlarged heads and expanded oral crushing surface, feed almost entirely on mussels and snails. The imported Asian Clam, *Corbicula fluminea,* now abounds in most of the streams within the turtle's

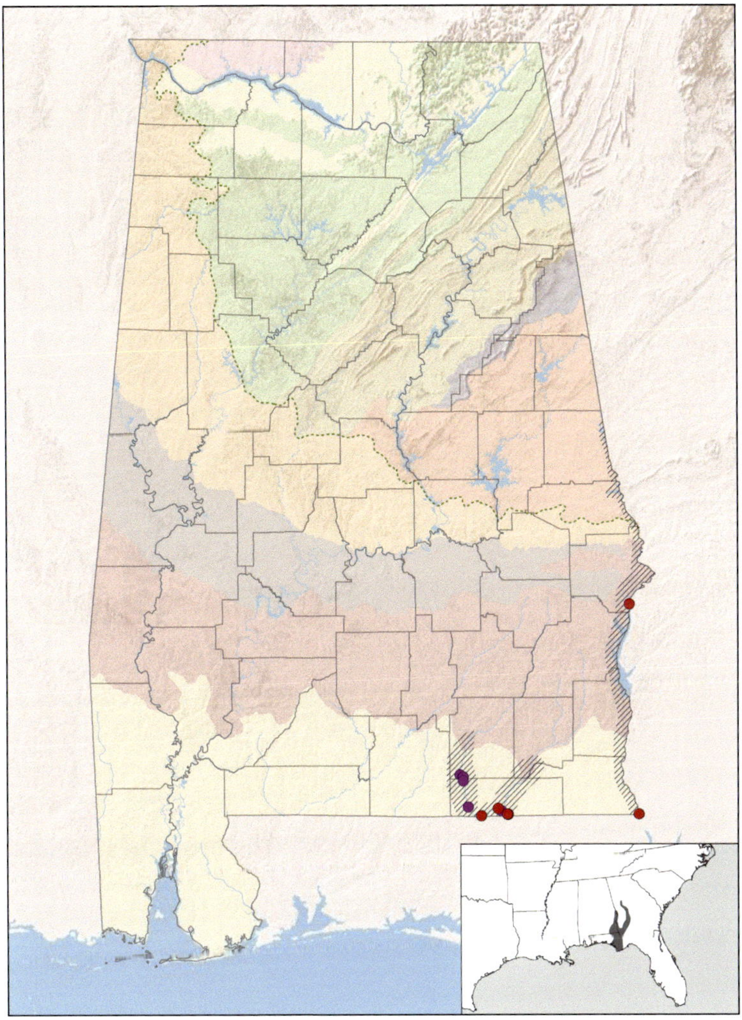

Alabama and continental US distribution of Barbour's Map Turtle; purple dots indicate hybrids with the Escambia Map Turtle

range, constituting an important food source. Females produce four to eleven eggs per clutch and are capable of developing two clutches each year. The eggs are laid in nests that are placed along major rivers and in chambers that are 3 to 6 inches (7.5–15 cm) deep, dug in moist sand of sandbars. Females migrate up and down rivers to use this limited resource (Sterrett 2009). The eggs are dull white and ellipsoidal in shape. Nesting occurs in early summer and hatching occurs in September, with hatchlings averaging about 1.5 inches (35 mm) in length.

Conservation and Management Barbour's Map Turtle is ranked Priority 2 (High Conservation Concern) by ADCNR (Mirarchi et al. 2004). For this reason, it is unlawful to possess this species in the state of Alabama without a special collecting permit. This turtle has suffered from overzealous collecting throughout its range and it is sold in the pet industry to European markets. Because of this persistent collection pressure it is not abundant anywhere. Additionally, the Chattahoochee River has several hydropower dams that alter stream flow and prevent free dispersal of the species, features that might prevent recovery of populations in this vital drainage. Finally, the conversion of forested landscapes to agriculture likely has decreased the abundance of this species by eliminating clear pools as well as basking and nesting sites (Sterrett et al. 2010), although its detection can be improved in some urban streams (Barrett and Guyer 2008). Remarkably, little research has been performed on the demography of this species and so information upon which to base management decisions is lacking. A population study of comparable quality to that of Shealy (1976) for *G. ernsti* is needed for *G. barbouri*. Pools of clear water with limestone shelving for hiding underwater and snags for basking above water are primary habitat features needed for the species. Protection of such sites should be a high priority within the range of the species. Additionally, protection of steep-sloped sand bars as nesting sites should improve conditions for Barbour's Map Turtle. Any conservation activities that preserve current stocks of native mussels or restores past mollusk beds will enhance populations of *G. barbouri*. The SWAP (Alabama Department of Conservation and Natural Resources, Division of Wildlife and Freshwater Fisheries 2005) recommends comprehensive surveys for this species every five to ten years to monitor its status within the state.

Taxonomy No subspecies are recognized within this species; it is most closely related to the broad-headed *Graptemys* (*G. ernsti* + *G. gibbonsi* + *G. pulchra*; Stephens and Wiens 2003).

Alabama Map Turtle

Graptemys pulchra (Baur, 1893)

Description This is a large aquatic turtle attaining a maximum carapace length of about 11 inches (275 mm) in females and about 5 inches (120 mm) in males. The carapace has a median keel accentuated by spines or prominent knobs on some of the vertebrals; these are most pronounced on the second and third vertebrals, becoming less conspicuous with age and virtually unnoticeable on large females. The posterior edge of the carapace is serrate, as it is in most members of the genus. The shell is not noticeably tall nor is the carapace humped in females, as it is in *G. ernsti*. The ground color of the carapace is dull green, with a median, longitudinal black stripe that may be broken in some individuals. The pleurals are unmarked or they have light reticular markings of varying intensity that follow the seams. Yellow concentric marks are present on the upper surfaces of each marginal, and wide concentric black markings that form semicircles are present on the undersurfaces of these scutes. Plastral markings on these turtles are usually limited to narrow dark lines following the seams. The back of the head of most individuals lacks the pair of oval light spots that are aligned with the paramedial stripes (those adjacent to the median stripe) of *G. ernsti*. The head does have a large yellow or greenish

Right: Adult female Alabama Map Turtle, Elmore County, AL

Previous: Hatchling Alabama Map Turtle, Elmore County, AL

yellow blotch between the eyes (interorbital) and a blotch of the same color on each side of the face behind the eyes (postorbital). The interorbital and postorbital blotches have a narrow connection behind the eyes and the interorbital blotch ends anteriorly in a blunt tip rather than the three narrow prongs of *G. ernsti*. The color pattern of the chin is variable, but usually involves a combination of interconnected longitudinal and laterally oriented light stripes or bars that form a V-shape anteriorly, a feature that distinguishes this species from *G. barbouri*, which possesses a U-shaped chin bar. The tail and limbs are striped. Sexual dimorphism is pronounced, with females attaining much greater sizes than males and developing greatly enlarged heads, as in *Graptemys barbouri* and *G. ernsti*.

Alabama Distribution *Graptemys pulchra* occurs in suitable habitats throughout the Mobile Bay drainage, except possibly for the Tallapoosa River system above the Fall Line. No *Graptemys* are found in the Perdido River, creating a divide between *G. pulchra*, in the Mobile drainage, and *G. ernsti*, in the Conecuh, Yellow, Choctawhatchee and Pea Rivers.

Habits The Alabama Map Turtle inhabits streams ranging in size from medium-sized creeks to large rivers. The primary requisite of a candidate stream of sufficient size appears to be an abundance of mollusks, especially mussels, on which the adult females depend almost

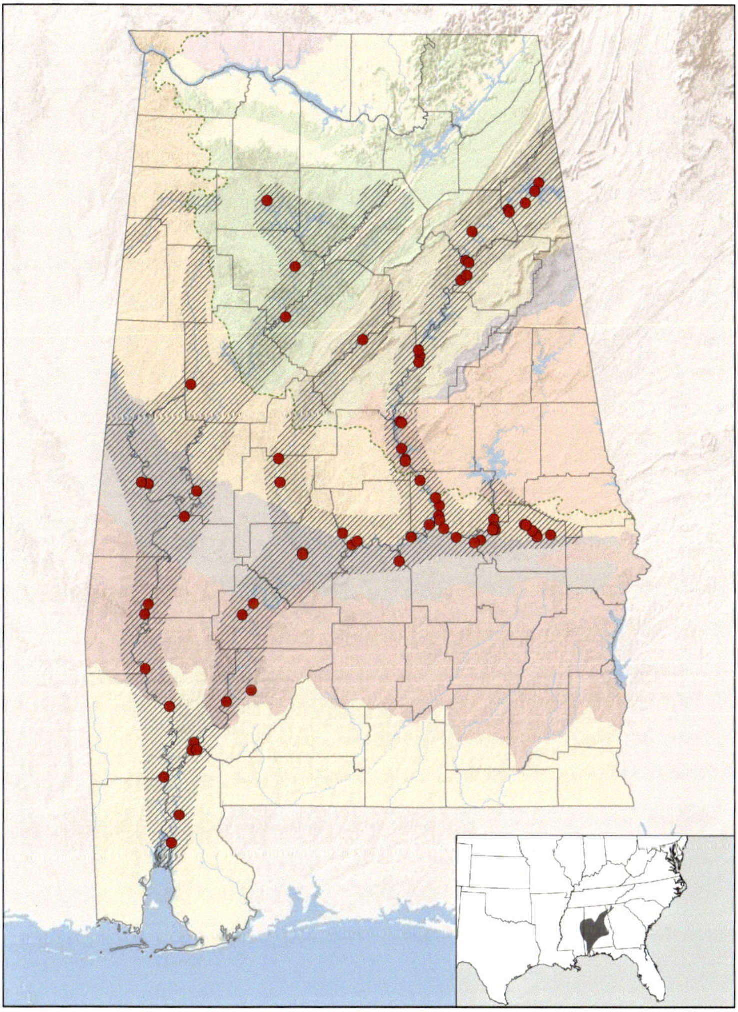

Alabama and continental US distribution of the Alabama Map Turtle

solely for food. Males and juvenile females feed mostly on insects and other arthropods.

Nesting activity in *G. pulchra* begins in late April or early May, reaches a peak in June, and continues through July and August. Up to seven clutches, averaging four to six eggs per clutch, may be laid by a female. Nest destruction by predators is heavy. The sex of offspring from these eggs is determined by the temperature of the nest, with females produced from warmer nests and males from cooler nests (Janzen and Paukstis 1991).

Conservation and Management Healthy populations of *G. pulchra* remain in the state. But this species is quite similar in appearance to the Escambia Map Turtle and both species are sold in the pet industry, especially in Europe. For these reasons both species receive protection by the state of Alabama, where it is unlawful to possess either species without a special collecting permit. Factors that contribute to a stream's capacity to support a sizable population of Alabama Map Turtles include the presence of nesting sites in the form of expansive sandbars and sandy banks, pools of water deep enough to provide shelter, and an abundance of snags, logs, and appropriately situated tree trunks to serve as basking sites, hiding places, and substrates for the attachment and development of food organisms. The practice of removing snags and logs from streams doubtless affects *G. pulchra* adversely, as it does most other stream-dwelling animals. Any conservation activities that preserve current stocks of native mussels or restores past mollusk beds will enhance populations of *G. pulchra*.

Taxonomy We follow Lovich and McCoy (1992) in separating *G. ernsti* from *G. pulchra*. This species is part of the broad-headed map turtle lineage, being most closely related to the common ancestor of *G. ernsti* + *G. gibbonsi* (Stephens and Wiens 2003). No subspecies are recognized within *G. pulchra*.

Escambia Map Turtle

Graptemys ernsti (Lovich and McCoy, 1992)

Description This highly aquatic species is large, attaining a maximum carapace length of about 11 inches (285 mm) in females and 5 inches (120 mm) in males. The carapace has a median keel that is accentuated by spines or prominent knobs on some of the vertebrals. These are most pronounced on the second and third vertebrals, becoming less conspicuous with age, and virtually unnoticeable on large females. The posterior border of the carapace is serrate. In profile, the carapace of females is unusually tall and humped, especially in individuals exceeding 3 inches (80 mm) carapace length. The ground color of the carapace is olive, with a median, longitudinal black stripe that is broken on some individuals. The pleural scutes have wide yellow rings along with light reticular markings of varying intensity that follow the seams. Yellow bar marks on the upper surface of each marginal and a dark smudge mark at the posterior edge of each marginal are visible when viewed from above. The plastral markings on these turtles are usually limited to narrow dark lines that follow the seams. In ventral view, the posterior border of each marginal has diffuse black marks that form semicircles on some scutes. On the back of the head is a pair of oval light spots that are aligned anterior to the paramedial stripes (those adjacent to the median stripe); these spots

Right: Juvenile Escambia Map Turtle, Covington County, AL

Previous: Adult female Escambia Map Turtle, Covington County, AL

are fused to paramedial stripes in some individuals of *G. ernsti,* a feature that is never found in *G. pulchra,* the most similar member of Alabama's *Graptemys* fauna. The head has one large yellow or greenish yellow blotch between the eyes (interorbital), and a large blotch of the same color on each side behind the eyes (postorbital). The postorbital and interorbital blotches are not interconnected (connected in *G. pulchra*), although they approach each other closely. The interorbital blotch typically ends in three narrow prongs that point toward the tip of the snout. The color pattern on the undersurface of the chin is variable, but it usually involves a combination of interconnected longitudinal and laterally oriented light stripes that form a V-shape (U-shaped in *G. barbouri*). The tail and the limbs are striped. Sexual dimorphism is pronounced in Escambia Map Turtles, with females attaining much greater sizes than males and developing greatly enlarged heads.

Alabama Distribution *Graptemys ernsti* occurs in flowing waters of the Yellow, Conecuh, and Choctawhatchee-Pea drainages, co-occurring with *G. barbouri* in the Pea River. Escambia Map Turtles are not known from the Perdido or Blackwater Rivers, probably because neither contains appropriate food, especially for females. This creates a dividing line between *G. pulchra,* in the Mobile drainage, and *G. ernsti.*

Habits The Escambia Map Turtle inhabits streams ranging in size from medium-sized creeks to large rivers, and the species may be the

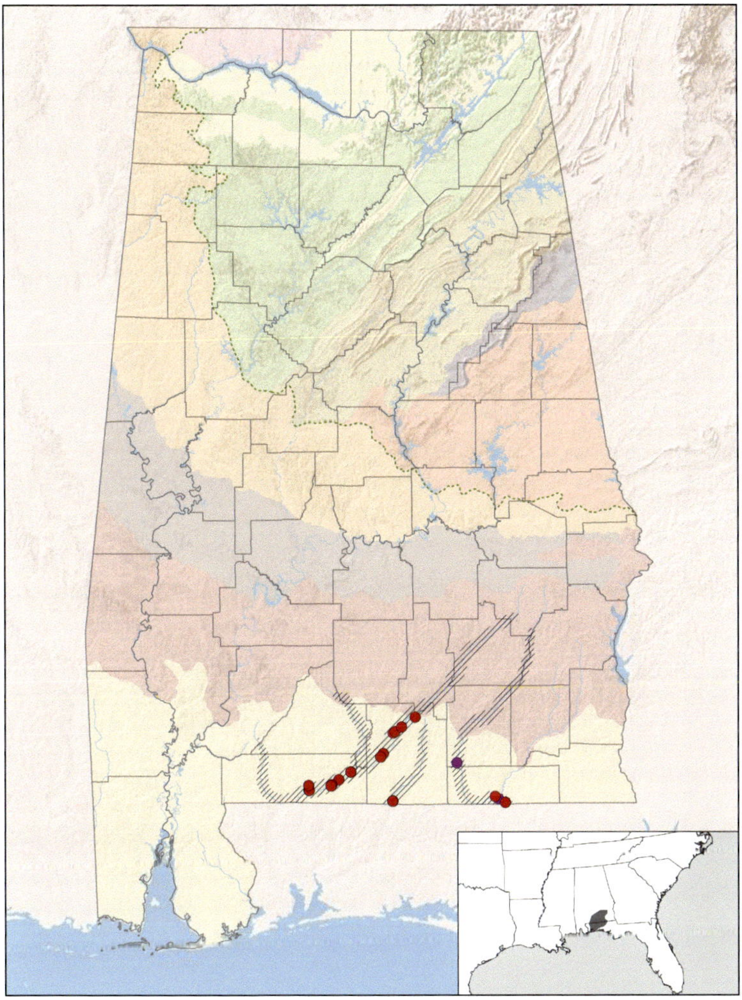

Alabama and continental US distribution of the Escambia Map Turtle; purple dots indicate hybrids with the Barbour's Map Turtle

most abundant turtle in the major river channels within its geographic range (Shealy 1976). This species basks on logs but, like all members of the genus, is extremely skittish and dives off these perches when approached by humans in boats. They can be captured by swimming with a piece of brush to hide behind and grabbing them on their basking sites. They can also be captured by diving and following them to their underwater roosts or by placing a dip net under their basking site. Females show great site fidelity to their basking logs, returning to them when displaced distances of 15 miles (24 km); males displaced similar distances have not been observed to return (Shealy 1976).

The primary requisite of a candidate stream of sufficient size appears to be an abundance of mollusks, especially mussels, on which the adult females depend almost solely for food. The imported Asian Clam, *Corbicula fluminea*, has apparently replaced native mollusks in rivers inhabited by *G. ernsti* and has become a primary diet item of these turtles. Males and juvenile females feed mostly on insects and other arthropods.

Sexual size dimorphism is pronounced in this species, with males being about half the size of females. Males possess mature sperm throughout the active season, and therefore seem capable of mating at any time. However, courtship behaviors are observed in September and November, when males approach the cloacal region of females, apparently to determine their receptivity (Shealy 1976). A male then swims in front of a receptive female and rapidly vibrates his head up and down along one side and then the other of the female's head (Shealy 1976). Males then mount the female from the rear and wrap their tails around the tails of receptive females during copulation. Females apparently store sperm from fall matings until ovulation the following April and May (Shealy 1976). Nesting activity in *G. ernsti* begins in late April or early May, reaches a peak in June, and continues through July and August. Up to seven clutches, each averaging four to six eggs, are laid by a female. Nests are constructed on large sand bars along meandering rivers and may be up to 65.5 feet (20 m) from the water's edge. Destruction of nests by predators is heavy. Shealy (1976) estimated that each year more than 90 percent of the nests are destroyed, with raccoons and fish crows being the most significant predators.

Conservation and Management The Escambia Map Turtle is ranked Priority 2 (High Conservation Concern) by ADCNR (2012 modification of Mirarchi et al. 2004). For this reason, it is unlawful to possess this species in the state of Alabama without a special collecting permit. This species is abundant in the Conecuh River and common in the Yellow River. Factors that contribute to a stream's capacity to support a sizable population of Escambia Map Turtles include the presence of nesting sites in the form of expansive sandbars and sandy banks, pools of water deep enough to provide shelter, and an abundance of snags, logs, and appropriately situated tree trunks to serve as basking sites, hiding places, and substrates for the attachment and development of food organisms. The practice of removing snags and logs from streams doubtless affects *G. ernsti* adversely, as it does most other stream-dwelling animals. Any conservation activities that preserve current stocks of native mussels or restores past mollusk beds will enhance populations of *G. ernsti*.

Taxonomy We follow Lovich and McCoy (1992) in separating *G. ernsti* from *G. pulchra*. No subspecies are recognized within *G. ernsti*. This species is the sister to *G. gibbonsi* (Stephens and Wiens 2003), the Pascagoula Map Turtle of the Pascagoula and Pearl Rivers of Mississippi.

Ouachita Map Turtle (False Map Turtle)

Graptemys ouachitensis ouachitensis (Cagle, 1953)

Description This is a fairly large aquatic turtle that attains a maximum carapace length of around 10 inches (255 mm) in females and 5.5 inches (140 mm) in males. The carapace has a median keel accentuated by prominent knobs or blunt spines on at least the second and third vertebrals. Along its posterior border, the carapace is serrate. The carapace has a ground color of brown or greenish olive with an often interrupted, black, mid-dorsal stripe and a network of yellowish lines, these usually forming circles or semicircles on the pleurals. The plastron is yellowish with a dark complex figure that becomes obscure with age. The undersurfaces of the marginals have concentric dark rings along the seams, and the bridge is marked with dark lines. This species has a head that is brownish or olive with yellow lines. Behind each eye is a rectangular-to-oval yellow spot with one to three yellow lines entering the orbit below this spot. Below each eye is an oval spot, and three yellow spots are found on the undersurface of the lower jaw, one on each side of the jaw and one on the tip of the chin. This triad of spots, along with the rectangular postorbital spot, are unique among Alabama's *Graptemys*.

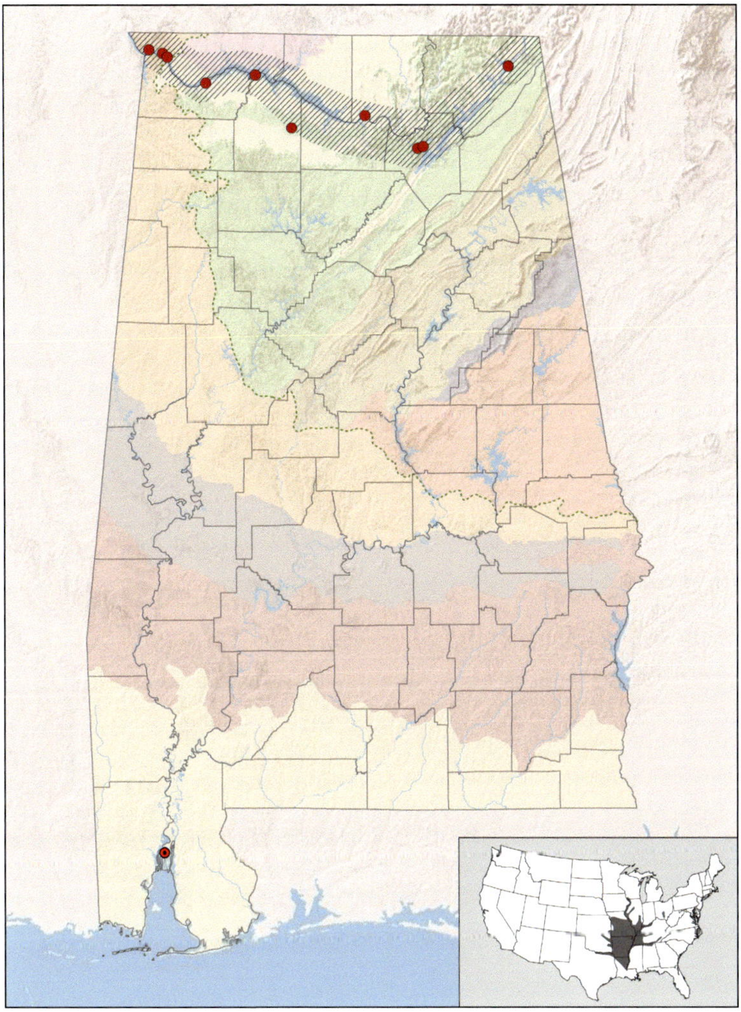

Left: Alabama and continental US distribution of the Ouachita Map Turtle

Opposite: Adult Ouachita Map Turtle, Alabama

Alabama Distribution In Alabama this species is generally confined to the Tennessee River and its associated backwaters. However, a single specimen collected in Baldwin County indicates that this species may have used the Tenn-Tom Waterway to expand its geographic range into the Mobile drainage.

Habits This turtle is essentially an inhabitant of large impoundments and low-gradient streams. The Northern Map Turtle, *Graptemys geographica*, with which it shares much of its range, is often found in small, shallow creeks, well upstream from their mouths, but *G. ouachitensis* seldom ascends the small tributaries.

This turtle, like the other *Graptemys*, is fond of basking and is frequently seen in considerable numbers on logs, snags, and stumps, especially during spring months. Courtship involves face-to-face confrontation in which the male vibrates the front toenails against the snout and face of the female. The nesting habits are not well known. The egg complement reportedly varies from seven to thirteen and is laid in a flask-shaped excavation in the soil. The sex of offspring from these eggs is determined by the temperature of the nest, with females produced from warmer nests and males from cooler nests (Janzen and Paukstis 1991).

The food habits of *G. ouachitensis* are not well known. The few reports available seem to indicate that the young are entirely carnivorous, while the adults supplement a predominantly carnivorous diet with some plant material.

Conservation and Management Because the Ouachita Map Turtle is restricted to a single drainage, it is tracked as vulnerable by the Alabama Natural Heritage Program. However, the species appears to be abundant in that drainage and has extensive populations outside of the state. Nevertheless, it is unlawful to harvest this species for profit. Up to two individuals per day may be possessed by an individual, but these must be for personal use only. Retention of sand banks for nesting and of snags as basking sites are the key management efforts required to maintain this species.

Black-knobbed Sawback Turtles

Graptemys nigrinoda (Cagle, 1954)

Taxonomy This species is restricted to the Alabama and Tombigbee Rivers of Alabama and Mississippi and is part of the narrow-headed radiation of Map Turtles (Stephens and Wiens 2003). Two subspecies are recognized within this group, both of which occur in Alabama.

Key to the Subspecies of *Graptemys nigrinoda* of Alabama

1a Plastron plain or with a pattern occupying less than 60 percent of its surface; light stripes on soft parts wider than dark ones.

***Graptemys nigrinoda nigrinoda*—Northern Black-knobbed Sawback . . . . page 196.**

1b Plastron with a pattern occupying 70 percent or more of its surface; dark stripes on soft parts wider than light ones.

***Graptemys nigrinoda delticola*—Southern Black-knobbed Sawback . . . . page 200.**

Northern Black-knobbed Sawback

Graptemys nigrinoda nigrinoda (Cagle, 1954)

Description This subspecies is highly aquatic and medium in size, attaining a maximum carapace length in females of at least 6 inches (155 mm) and in males of about 4.5 inches (115 mm). The carapace has a medial knob-like projection on the second and third vertebrals, and a smaller one on the first vertebral. These projections become reduced to small, pointed swellings in old females. Along its posterior border, the carapace is serrate. Greenish olive or brown are the dominant colors of the carapace, and the medial projections are jet black. The pleurals are marked with yellow or white circles, and the sides of the vertebrals have light reticular markings. The plastron is yellowish with dark lines bordering the sutures of the scutes or it has a dark figure occupying not more than 30 percent of the plastral area. The undersurfaces of the marginals have concentric dark rings at the seams. In this subspecies, the head of both males and females is small, olive to brown, and lined with yellowish stripes. The head also possesses a pair of crescent-shaped, strongly recurved, postocular yellowish marks

Above: Adult female Northern Black-knobbed Sawback, Elmore County, AL

Opposite: Hatchling Northern Black-knobbed Sawback, Elmore County, AL

that usually are connected posteriorly to longitudinal stripes on the neck. In overall appearance this subspecies has skin that is light in color and a shell that is flatter in shape than *G. n. delticola*.

Alabama Distribution This subspecies is found in the Coosa, Tallapoosa, and Cahaba Rivers from the Fall Line southward, and the Alabama River downstream to approximately the level of the Wilcox-Monroe County line, where integradation with *G. n. delticola* begins. Folkerts and Mount (1969) considered the northernmost populations of *G. nigrinoda* in the Warrior River to be of the nominate subspecies, but these are now considered to be intergradient. Populations farther south in the Warrior River are also intergradient, as are those in the Tombigbee River southward at least to the city of Jackson, Clarke County.

Habits These turtles are found in rivers and seem never to leave the water except to lay eggs or to bask, the latter being a habit to which they are strongly addicted. This subspecies is less abundant than the southern subspecies and is not as dominant a member of the turtle

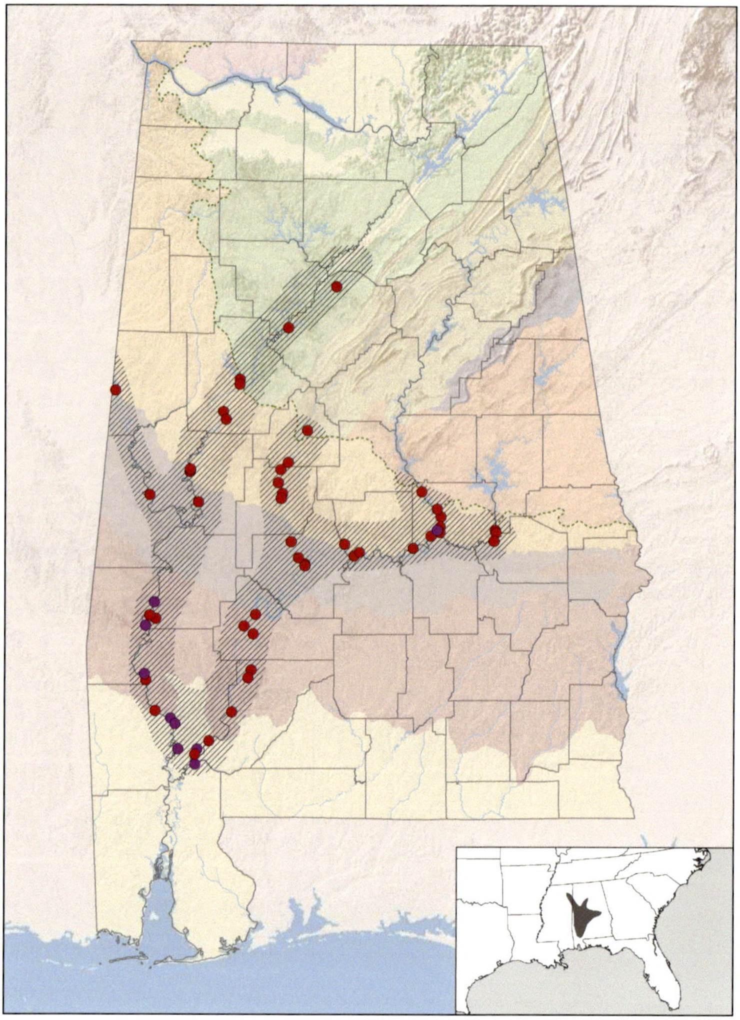

Alabama and continental US distribution of the Northern Black-knobbed Sawback; purple dots indicate intergrades with the Southern Black-knobbed Sawback

fauna where it occurs. Like all members of the species, this subspecies is carnivorous, but details of its food habits are poorly known. Insect remains have been observed in feces of freshly captured specimens from Alabama. Data on reproduction are scarce, but specimens from Alabama lay up to six oblong eggs in July.

Conservation and Management The Northern Black-knobbed Sawback is ranked Priority 3 (Moderate Conservation Concern) by ADCNR (Mirarchi et al. 2004). For this reason, it is unlawful to possess this species in the state of Alabama without a special collecting permit. Healthy populations of this subspecies remain in the state. However, it is sold in the pet trade, especially in Europe, a feature that may create a black market within Alabama. Retention of sand banks for nesting and of snags as basking sites are the key management efforts required to maintain this subspecies.

Taxonomy Ennen et al. (2014) examined this subspecies and the Southern Black-knobbed Sawback and concluded that the two represent clinal variation rather than distinctive evolutionary units. They suggest that the lighter coloration and flatter body shape of Northern Black-knobbed Sawbacks indicate adaptation to the flowing clear waters occupied by that subspecies. We retain the subspecific designations because we expect subspecies to exhibit the levels of gene flow coupled with distinctive morphology and coloration documented in their study.

Southern Black-knobbed Sawback

Graptemys nigrinoda delticola (Folkerts and Mount, 1969)

Description Southern Black-knobbed Sawbacks are highly aquatic and medium in size, attaining a maximum carapace length of 8 inches (210 mm) in females and about 4.5 inches (115 mm) in males. The carapace has a medial knob-like projection on the second and third vertebrals, and a smaller one on the first vertebral. These projections become reduced to small, pointed swellings in old females. Along its posterior border, the carapace is serrate. Dark brown or black are the dominant colors of the carapace. The plastron pattern is complex, with dark lines and rounded figures bordering the sutures and covering more than 60 percent of the plastral area. In this subspecies, the head of both males and females has postorbital marks that are neither crescent-shaped nor strongly recurved laterally; these marks often are disconnected from the longitudinal stripes on the head. The skin of this subspecies is predominantly dark, usually black, with narrow light stripes. In general appearance this subspecies is darker and has a greater shell height than the nominate subspecies.

Alabama Distribution Southern Black-knobbed Sawbacks are confined to Alabama, where they are found in the Mobile, Tensaw, and

Left: Adult male Southern Black-knobbed Sawback, Baldwin County, AL

Opposite: Adult female Southern Black-knobbed Sawback, Baldwin County, AL

other anastomosing freshwater streams and lakes that constitute the upper portion of the delta of the Mobile Bay drainage, in Baldwin and Mobile Counties.

Habits In Coastal Plain stretches of the large rivers that drain into Mobile Bay, *G. n. delticola* occurs in abundances that are so great that this subspecies dominates the local turtle assemblages. The diet is quite varied and includes both animal and algae. Animal materials in the diet are dominated by sessile invertebrates (sponges and bryozoans) and algal materials are dominated by *Cladophora*, *Ulothrix*, and *Spirogyra*.

Reproductive data for this species are available from the extensive population study of Lahanas (1982) for individuals on Gravine Island. At this site, nesting starts in late May, reaches a peak in late June and early July, and ends early in August. Females deposit eggs only at night, and, because this subspecies is so abundant, ovipositing females assemble off of nesting beaches at dusk. After nightfall, these females search for areas of dry sand 6.5–13 feet (2–4 m) above the water line where they dig an urn-shaped nest, deposit three to seven eggs, cover the nest, and then return to the water. The eggs are smooth, pink, elliptical, and have a soft shell. Like nesting female sea turtles, Southern Black-knobbed Sawback females are extremely

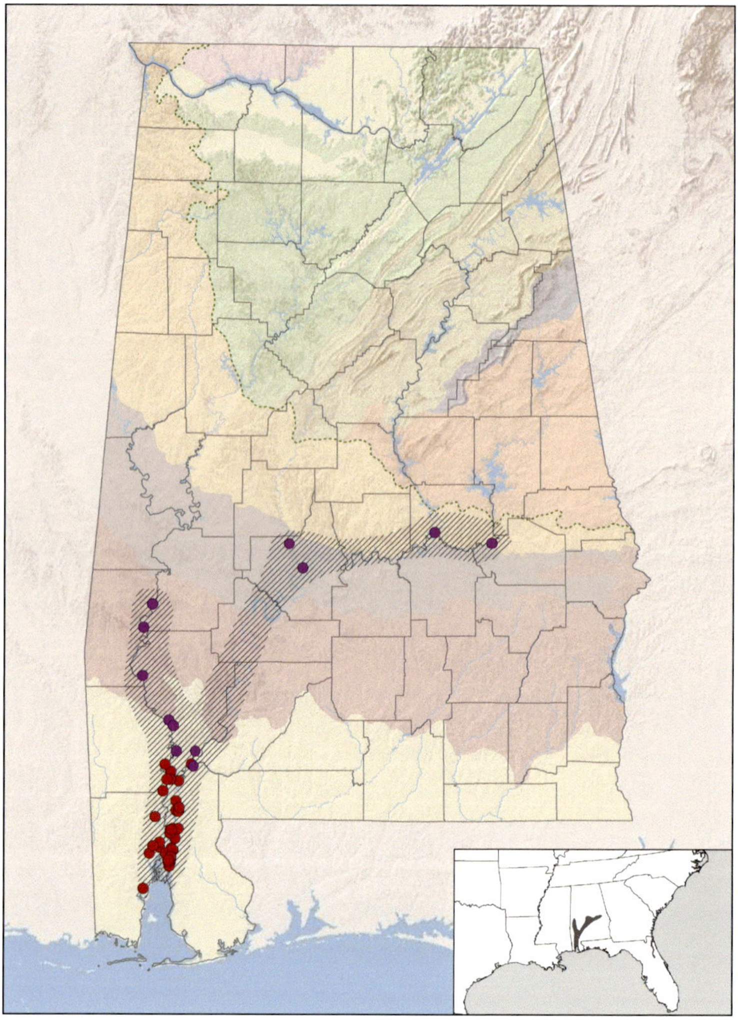

Alabama and continental US distribution of the Southern Black-knobbed Sawback; purple dots indicate intergrades with the Northern Black-knobbed Sawback

skittish and will break off nesting attempts with minimal provocation. Females are thought to be capable of depositing up to three clutches per season. Hatchlings emerge from mid-August to early October. Individuals from a clutch of eggs hatch asynchronously, emerging over a period of up to three days.

Conservation and Management The Southern Black-knobbed Sawback is ranked Priority 3 (Moderate Conservation Concern) by ADCNR (Mirarchi et al. 2004). For this reason, it is unlawful to possess this subspecies in the state of Alabama without a special collecting permit. These turtles remain an abundant member of turtle assemblages in the Mobile, Tensaw, and lower Alabama Rivers. However, it is prized in the pet trade, especially in Europe, perhaps creating a black market within Alabama. Retention of sand banks for nesting and of snags as basking sites are the key management efforts required to maintain this subspecies.

TAXONOMY Ennen et al. (2014) examined this subspecies and the Northern Black-knobbed Sawback and concluded that the two represent clinal variation rather than distinctive evolutionary units. They suggest that the darker coloration and domed body shape of Southern Black-knobbed Sawbacks indicate adaptation to the slow-current turbid waters occupied by this subspecies. We retain the subspecific designations because we expect subspecies to exhibit the levels of gene flow coupled with distinctive morphology and coloration documented in their study.

Sliders

Genus *Trachemys* (Agassiz, 1857)

We follow Seidel (2002) in considering this genus to contain fifteen species of freshwater turtles from North, Central, and South America as well as from the Greater Antilles. Unfortunately, members of this genus are a paraphyletic assemblage distributed along a branch leading from a lineage creating a monophyletic genus *Pseudemys* to the common ancestor of *Graptemys* + *Malaclemys* (Stephens and Wiens 2003). Therefore, membership within this genus is likely to be modified by future revisions. Fortunately, for the purposes of taxa in Alabama, the type specimen is *Trachemys scripta* and this is the species that is found in the state. Therefore, any name changes associated with taxonomic revisions should not affect the genus name used for Alabama specimens.

Pond Sliders

Trachemys scripta (Schoepff, 1792)

Taxonomy This species, whose range extends from coast to coast and from the Great Lakes southward to Florida, is the most commonly observed turtle in the United States. Three subspecies traditionally are recognized and these form a monophyletic cluster at the base of the paraphyletic assemblage of species currently placed in the genus *Trachemys* (Seidel 2002; Stephens and Wiens 2003). Two of the three subspecies, *T. s. elegans* and *T. s. scripta,* occur in Alabama. Specimens with characteristics similar to *T. s. troosti* are found in northeast Alabama, but we follow Davidson (1971) in interpreting these to be intergradient between *T. s. elegans* and *T. s. scripta*. Previous authors have placed this species in the genera *Chrysemys, Emys, Pseudemys,* and *Testudo.*

Key to the Subspecies of *Trachemys scripta* of Alabama

1a Plastron with twelve dark spots or smudges, each in center of plastral scutes; head with an orange or reddish temporal bar; stripes on pleural scutes relatively narrow.

Trachemys scripta elegans—**Red-eared Slider** . . . . **page 206.**

1b Plastron with dark smudges or spots on gular scutes; head with a yellow blotch on side; stripes on pleural scutes relatively wide

Trachemys scripta scripta—**Yellow-bellied Slider** . . . . **page 209.**

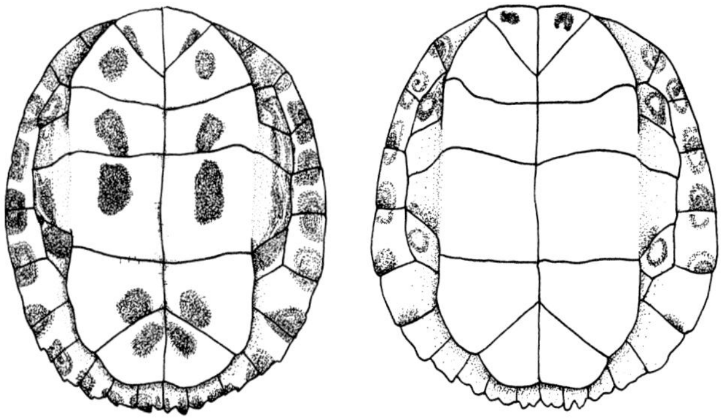

From left to right:

Plastron of Red-eared Slider with twelve dark spots

Plastron of Yellow-bellied Slider with dark spots on gular scutes

Red-eared Slider

Trachemys scripta elegans (Wied-Neuwied, 1838)

Description This subspecies is a moderately large, robust, aquatic turtle that attains a maximum carapace length of about 11.5 inches (275 mm). The carapace is rounded to slightly oval in shape, has a relatively low dome, becomes longitudinally wrinkled in old individuals, and has a ground color that is olive green to black with a narrow, vertical yellow stripe on each pleural scute (shell with high dome, deeper wrinkles, and broader yellow stripes in *P. s. scripta*). The head has an orange to red longitudinal bar located behind each eye (a large yellow blotch characterizes *P. s. scripta*), and the chin and neck have broad stripes that are relatively few in number. The iris has a transverse dark bar and the upper jaw has a terminal notch, but this is not flanked by projecting cusps (as in *P. alabamensis*). The plastron is large and rigid and typically has a large dark spot or smudge at the center of each scute (these sharply defined and frequently ocellate in young; dark spots only on gular in *P. s. scripta*). The bridge and undersurfaces of each marginal are more black than yellow. When viewed from behind, the rump typically has a series of longitudinal yellow stripes. Color patterns of adults become obscure with age, especially in males who may become melanistic over time. The ground color of hatchlings is greenish.

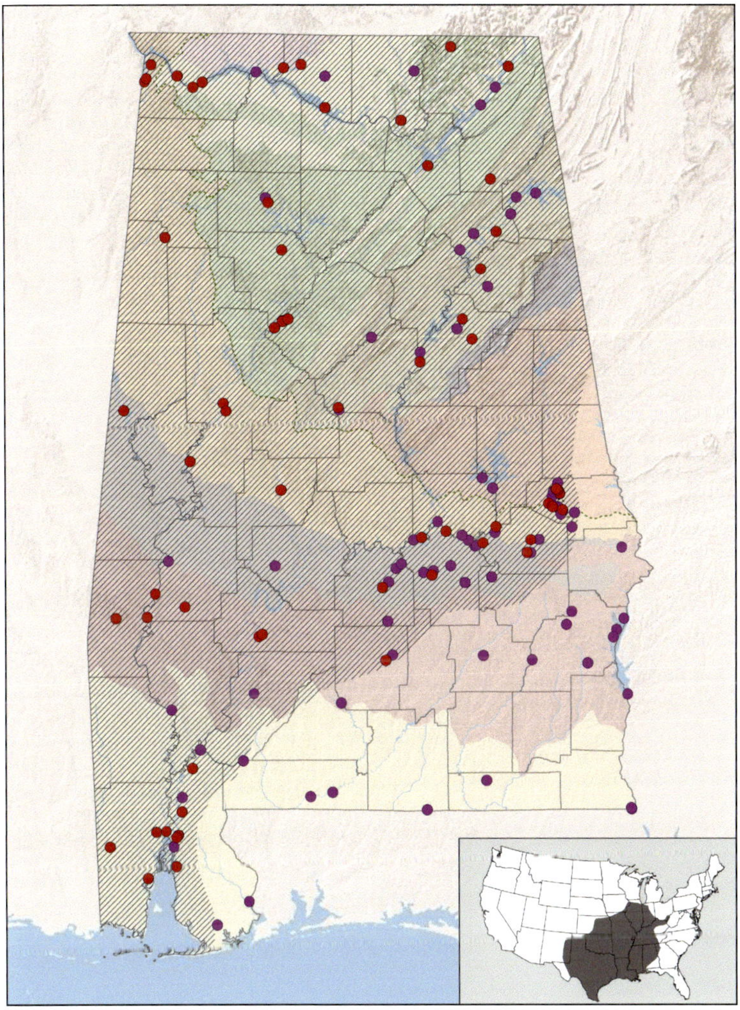

Left: Alabama and continental US distribution of the Red-eared Slider; purple dots indicate intergrades with the Yellow-bellied Slider

Opposite: Juvenile Red-eared Slider (left) and Yellow-bellied Slider (right), Covington County, AL

Alabama Distribution Specimens referable to this subspecies are found in the Tennessee River and the Alabama River system including the Upland areas of Alabama. The influence of this subspecies is observed throughout the rest of the state, but becomes restricted to specimens from major river channels for rivers east of Mobile Bay (Davidson 1971).

Habits *Trachemys scripta elegans* inhabits a wide variety of permanently aquatic habits, but is generally most abundant in lakes and sluggish rivers. Apparently, it is better adapted to life in larger streams than *T. s. scripta,* which could account in part for its strong influence in and around the rivers that flow through the Red Hills and Lower Coastal Plain regions otherwise inhabited by populations whose characteristics are predominantly or exclusively those of *T. s. scripta.*

The eggs of Red-eared Sliders, varying in number from four to twenty per clutch, are laid in cavities dug by the females, often at considerable distances from the water. Up to three clutches may be laid in one season. Most of the hatchling turtles overwinter in the nest. The food of *T. s. elegans* consists chiefly of aquatic animal life, supplemented with vegetation.

Conservation and Management This is an exceptionally common subspecies that does well in farm ponds, urban parks, and golf course lakes. Nevertheless, populations can be overharvested and so this subspecies cannot be collected in Alabama for sale. Up to two individuals may be possessed per day for personal use. No special conservation efforts are required to maintain populations in its native range. However, because members of this subspecies, especially the females, cross roads frequently (Steen et al. 2006), wildlife fencing and designed underpasses in areas of heavy traffic may be necessary to maintain populations in urban settings. Hatchlings are produced commercially in some states, including Alabama, and provide the basis for most of the baby turtle trade in this country. Its introduction to areas outside of its native range is a growing problem.

Yellow-bellied Slider

Trachemys scripta scripta (Schoepff, 1792)

Description This subspecies is a moderately large, robust, aquatic turtle that attains a maximum carapace length of about 11.5 inches (275 mm). The carapace is rounded to slightly oval in shape, has a relatively high dome, becomes longitudinally wrinkled in old individuals, and has a ground color that is olive green to black with a broad, vertical yellow stripe on each pleural scute (shell with low dome, shallower wrinkles, and narrower yellow stripes in *P. s. elegans*). The head has a large yellow blotch behind each eye, and the chin and neck have broad stripes that are relatively few in number (red or orange blotch behind eye in *P. s. elegans*). The iris has a transverse dark bar and the upper jaw has a terminal notch, but this is not flanked by projecting cusps (as in *P. alabamensis*). The plastron is large and rigid and typically has two spots, one on each gular scute (dark spot on all plastral scutes in *P. s. elegans*). The bridge may have small spots or may be unmarked. When viewed from below, each marginal has a dark spot that is smaller in area than the yellow ground color. When viewed from behind, the rump typically has a series of longitudinal yellow stripes. Color patterns of adults become obscure with age, especially in males who may become melanistic over time. The ground color of hatchlings is greenish.

Right: Adult male Yellow-bellied Slider, Jefferson County, FL

Previous: Adult female Yellow-bellied Slider, Covington County, AL

Alabama Distribution Influence of this subspecies occurs throughout the state. However, populations most strongly referable to *T. s. scripta* occur in the Lower Coastal Plain east of Mobile Bay (Conecuh, Yellow, Choctawhatchee, Pea, and Chattahoochee drainages), where this subspecies is associated with ponds, swamps, and lakes (Davidson 1971). Populations that are intergradient between *T. s. scripta* and *T. s. elegans* inhabit all drainages within Alabama.

Habits This turtle seems best adapted to aquatic environments with abundant vegetation. Within its range, such habitats include lakes, ponds, swamps, sloughs, and low-gradient, shallow streams. Overland migrations are common in *T. s. scripta,* accounting for its frequent occurrence in semi-permanent bodies of water.

In the first year of growth, young Yellow-bellied Sliders are rather strongly carnivorous, relying heavily on aquatic insects and other invertebrate life for food. Older ones are mostly herbivorous, including in their diet a wide variety of aquatic plants. Many farm pond owners assume that Yellow-bellied Sliders are harmful to fish populations and, therefore, destroy the turtles, but there is no evidence that turtles influence size or abundance of fish.

Eggs, usually between ten and fifteen per clutch, are laid in a flask-shaped excavation, typically within 330 feet (100 m) of water. The

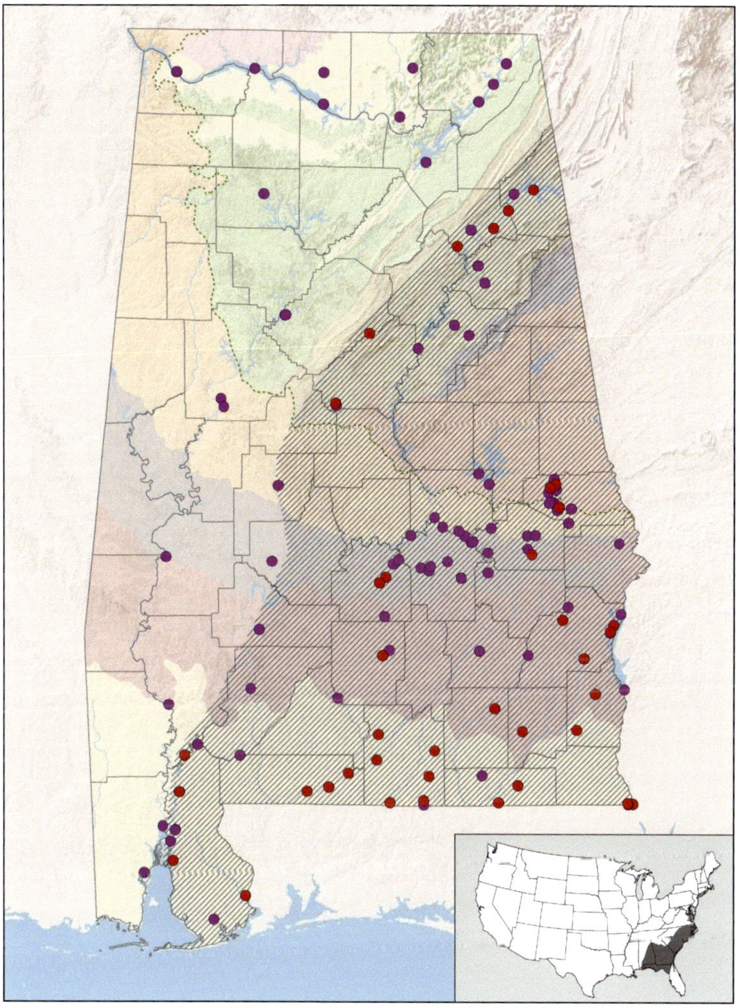

Alabama and continental US distribution of the Yellow-bellied Slider; purple dots indicate intergrades with the Red-eared Slider

largest numbers of hatchlings appear in spring, indicating that many overwinter in the nest.

Adults of this subspecies are rather ill-tempered and are more inclined to bite than our other river and pond turtles, but they do rather well in captivity.

Conservation and Management This is a common subspecies that does well in farm ponds, urban parks, and golf course lakes. However, populations can decline from overharvesting and so these turtles cannot be collected for sale. Up to two individuals may be possessed per day for personal use. In rivers and large streams these turtles increase in abundance and dominate turtle assemblages in areas where the forested landscape has been converted to agriculture (Sterrett et al. 2010). No special conservation efforts are required to maintain populations in its native range. However, because members of this subspecies, especially the females, cross roads frequently (Steen et al. 2006), wildlife fencing and designed underpasses in areas of heavy traffic may be necessary to maintain populations in urban settings.

Cooters

Genus *Pseudemys* (Gray, 1856)

This is the genus of the Red-bellied Turtles and Cooters, a radiation that occupies rivers, lakes, and ponds. Seven species are contained within this genus, all of which are largely herbivorous and whose aggregate geographic range includes the eastern United States (Stephens and Wiens 2003). These turtles are most similar to *Trachemys* in general size, shape, and color, but differ from that genus in lacking dark smudge marks on the center of at least the anteriormost plastral scutes. These turtles are the sister taxon to the ancestor of *Trachemys* + *Graptemys* + *Malaclemys* (Stephens and Wiens 2003). Three species occur in Alabama.

Key to the Species of *Pseudemys* of Alabama

1a Upper jaw with a terminal notch, flanked on each side by a cusp.

Pseudemys alabamensis—**Alabama Red-bellied Turtle . . . . page 216.**

1b Upper jaw with or without a terminal notch; terminal notch, if present, not flanked, on each side by a cusp. Go to **2**.

Head of Alabama Red-bellied Turtle with terminal notch and cusp on upper jaw

2a Undersides of all posterior marginals with at least some dark pigment; plastron usually with some dark markings; second pleural scute of carapace with a C-shaped light figure; iris of eye entire in many populations.

Pseudemys concinna—**River Cooter . . . . page 221.**

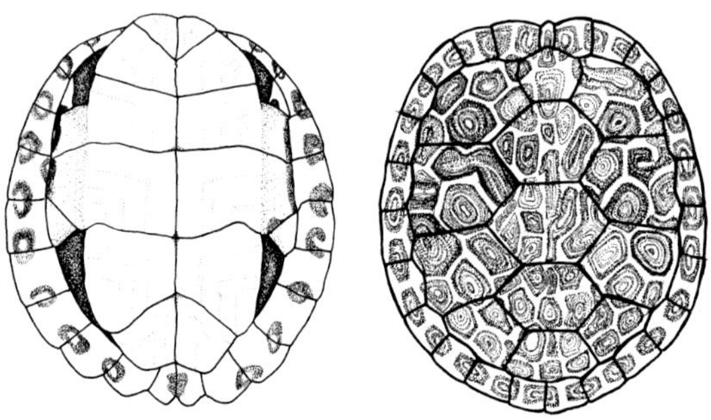

From left to right:

Plastron of River Cooter with some dark markings

Carapace of River Cooter with a C-shaped light figure

2b Undersides of most posterior marginals lacking dark pigment (Alabama populations); plastron unmarked (except in a few old adults); second pleural scute of carapace usually with a vertical mark, which may have upper or lower bifurcations; iris of eye bisected (or nearly bisected) by a transverse dark bar.

Pseudemys floridana—**Florida Cooter . . . . page 226.**

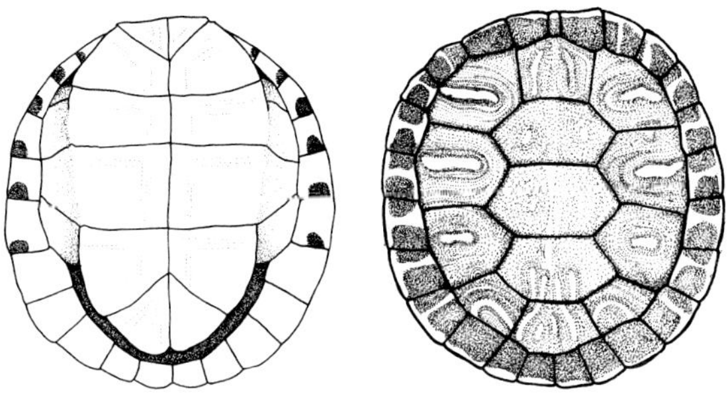

From left to right:

Plastron of Florida Cooter without dark markings

Carapace of Florida Cooter with vertical markings

Alabama Red-bellied Turtle

Pseudemys alabamensis (Baur, 1893)

Description These are large, robust turtles, attaining a maximum carapace length of about 13 inches (335 mm) in females and 11.5 inches (295 mm) in males. The shell is deep, with a carapace that is oval in shape and slightly serrate along its posterior border. The surface of the carapace becomes wrinkled longitudinally with age. The plastron is large with a deep bridge. In color, the carapace is olive to dark brown, with variable light markings that usually include a light vertical bar on the second pleural scute. The plastron is reddish, occasionally plain, but typically with an elaborate figure of dark bars and variably shaped, light-centered dark markings that may be separate or connected. The plastron and, less frequently, the carapace are freckled or marked with brownish vermiculations in some large individuals. The bridge is plain or with diagonally oriented dark markings. In ventral view, the marginals have a light-centered single dark ring or double dark rings. The skin is olive to black with yellow striping and the iris has a transverse dark bar. The key feature of the species is a terminal notch with a distinct cusp on each side at the front of the upper jaw. The notch accommodates a tall, pointed, toothlike projection from the tip of the lower jaw, and the cusps of the upper jar fit into notches to each side of the toothlike projection. These features are found in no other member

Left: Dorsal (upper) and ventral (lower) view of hatchling Alabama Red-bellied Turtle, Baldwin County, AL

Opposite: Adult female Alabama Red-bellied Turtle, Baldwin County, AL

of the family in Alabama. Hatchlings are greenish, have a conspicuously marked carapace, and resemble juveniles of *P. floridana,* from which they can be distinguished by the presence of the maxillary cusps and bright red bellies.

Alabama Distribution This species is found in waterways associated with the lower portion of the Mobile Bay drainage and the lower portion of the Pascagoula River and Back Bay of the Biloxi watershed (Leary et al. 2008). In Mobile Bay itself the turtle is found chiefly

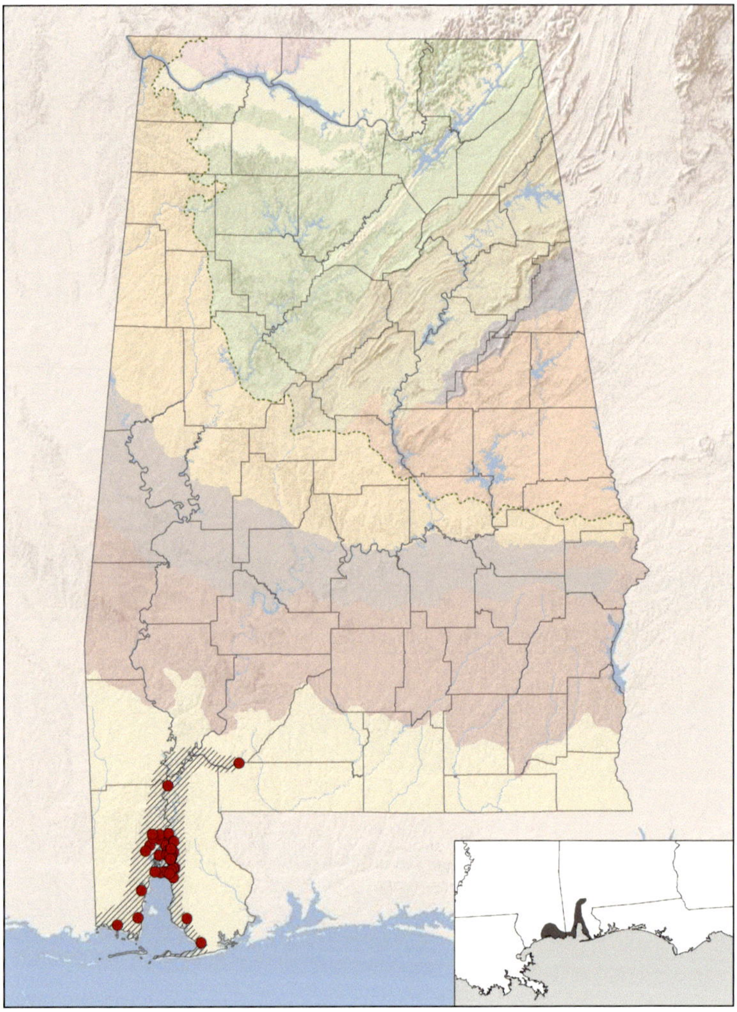

Alabama and Gulf Coast distribution of the Alabama Red-bellied Turtle

in the uppermost portion. The northern limit of the range is poorly known, but is documented to extend at least to Little River State Park Lake, in southern Monroe County.

Habits Optimum habitat conditions for Alabama Red-bellied Turtles are found in the stretch of the Tensaw River between Hurricane and Spanish Fort in Baldwin County. Although the species occurs sparingly in and around Mobile Bay and is captured occasionally on Dauphin Island, perhaps as a waif, it is not a salt marsh form. The species is most abundant in fresh to moderately brackish water that supports an abundance of submergent aquatic vegetation, the turtle's principal food. Preferred diet items include *Hydrilla, Najas, Vallisneria,* and *Sagitteria* (Leary et al. 2008). Individual turtles may move as much as 11 miles (17.9 km) during a season of activity. The species is active year round, but becomes dormant for short periods of time during the winter when cold fronts move through southern Alabama.

Alabama Red-bellied Turtles are fond of basking. The species nests on sand bars and the banks of large rivers. Nesting occurs as early as late April and as late as mid-August. During this time females seek sandy soils in the form of sand bars, dredge spoils, pine sandhills adjacent to rivers, or alligator nests. The nests tend to be created in late afternoon or evening hours and contain clutches of seven to nineteen eggs with an average of twelve (Nelson et al. 2009). In Alabama, major nesting sites include Gravine Island, a human-made island of dredge deposits, the Mobile Bay Causeway, Meaher State Park, Howard's Island, and Big Island. Most hatching occurs during September through November, but eggs in a few nests are thought to overwinter, with hatchlings emerging the following March or April (Nelson et al. 2009).

Conservation and Management The Alabama Red-bellied Turtle, the official state reptile of Alabama, is listed as endangered under the ESA. The species is also ranked Priority 1 (Highest Conservation Concern) by ADCNR (Mirarchi et al. 2004) and, by state statute, it is unlawful to possess this species without state and federal permits. Recent capture-recapture data suggest that a stable population occurs in Mobile Bay, but a reduced proportion of juveniles in recent samples suggests high nest and hatchling predation. Feral pigs, raccoons, fish crows, and fire ants are major predators that restrict recruitment of juveniles. In addition to juvenile mortality, the geographic range is not as extensive as it has been in the past, with no individual being discovered north of David Lake during recent surveys. Use of Gravine Island for recreation creates problems because of off-road-vehicle traffic at nesting beaches, adult mortality from boat propellers, and direct consumption of eggs or adults. Adult turtles also are killed in gill nets, hoop nets, and crab traps and are crushed while crossing roads, especially the US Highway 90 Mobile Bay causeway.

Nelson et al. (2009) demonstrated that eggs collected in the field could be hatched, reared, and released as a conservation tool to improve recruitment of juveniles. In that study, 175 hatchlings were released into the Blakeley River. A second conservation project has created a drift fence along the Highway 90 causeway to block turtles from access to the road. This project was designed initially to demonstrate the utility of fences in reducing adult mortality and consisted of 1,400 feet (427 m) of temporary fence constructed by volunteers. Following the apparent success of that effort, the Alabama Department of Transportation added 3.4 miles (5.5 km) of permanent fence that appears to have reduced road mortality along the Highway 90 causeway by as much as 80 percent. However, remaining turtles include many females seeking nest sites in loose dry soils along the causeway, which suggests that nesting habitat must be improved elsewhere along the causeway. The SWAP (Alabama Department of Conservation and Natural Resources, Division of Wildlife and Freshwater Fisheries 2005) recommends comprehensive surveys for this species every five to ten years to monitor its status within the state.

Taxonomy This species belongs to a Red-bellied Turtle group (Mount 1975), being most closely related to the common ancestor of *P. rubriventris* + *P. nelsoni* (Stephens and Wiens 2003). No subspecific variation is recognized within this species and previous authors have placed it in the genus *Chrysemys*.

River Cooter

Pseudemys concinna (Le Conte, 1830)

Description River Cooters are large aquatic turtles that attain a maximum carapace length of about 15.5 inches (395 mm) in females and 12.5 inches (320 mm) in males (Aresco and Dobie 2000). The carapace is oblong to oval and is somewhat serrate and flaring along its posterior boundary (especially in males). In ground color, the carapace is olive-brown to nearly black, with concentric light markings on the pleural scutes and usually with a C-shaped light figure on the second pleural, a feature that differentiates this species from *C. floridana* (light I- or Y-shaped mark on second pleural). The plastron is yellow to orange in ground color, occasionally being plain, but more often with dark pigmentation along at least some of the seams. These colors become obscure or disappear entirely with age (especially in females). On their undersurfaces, the marginals, including the hindmost ones, have light-centered dark spots or rings. The bridge possesses dark spots or longitudinal markings as well. The color of the head, limbs, and tail is olive-brown to black with greenish yellow to reddish orange stripes. The cutting edge of the upper jaw may be smooth or serrate, but Alabama specimens lack the two anterior subterminal cusps characteristic of *P. alabamensis*. The color pattern of the iris differs regionally within the state, with some areas having a transverse dark bar and other areas lacking this feature. First-year juveniles have

Right: Juvenile River Cooter, Bibb County, AL

Previous: Adult male River Cooter, Elmore County, AL

carapaces that are light olive to greenish with vivid light markings. In this same group the plastral markings, if present, are conspicuous and well defined.

Alabama Distribution This species is found throughout the state.

Habits The River Cooter is chiefly a resident of streams and relatively large lakes. It is often seen basking on logs or rocks and under such circumstances is exceedingly wary, plunging into the water at the slightest indication of human presence. Mature females are particularly difficult to approach. River Cooters are strongly aquatic and, unlike *T. scripta,* a species with which they often occur, seldom venture onto land except to nest. However, within their aquatic habitat they move freely and, occasionally, over long distances in their daily and weekly movements (less than 3.7 miles [under 6 km]; Sterrett et al. 2008). In Alabama, this species has a low, flat shell in populations inhabiting streams above the Fall Line and high, arched shells in populations of the Lower Coastal Plain (Aresco and Dobie 2000). This change in morphology may be related to the faster-flowing waters above the Fall Line and the increased pressure of alligator predators in the Lower Coastal Plain.

Our description of reproduction is based on extensive data collected by Ken Fahey (1987) at Horseshoe Bend in Tallapoosa County. Males

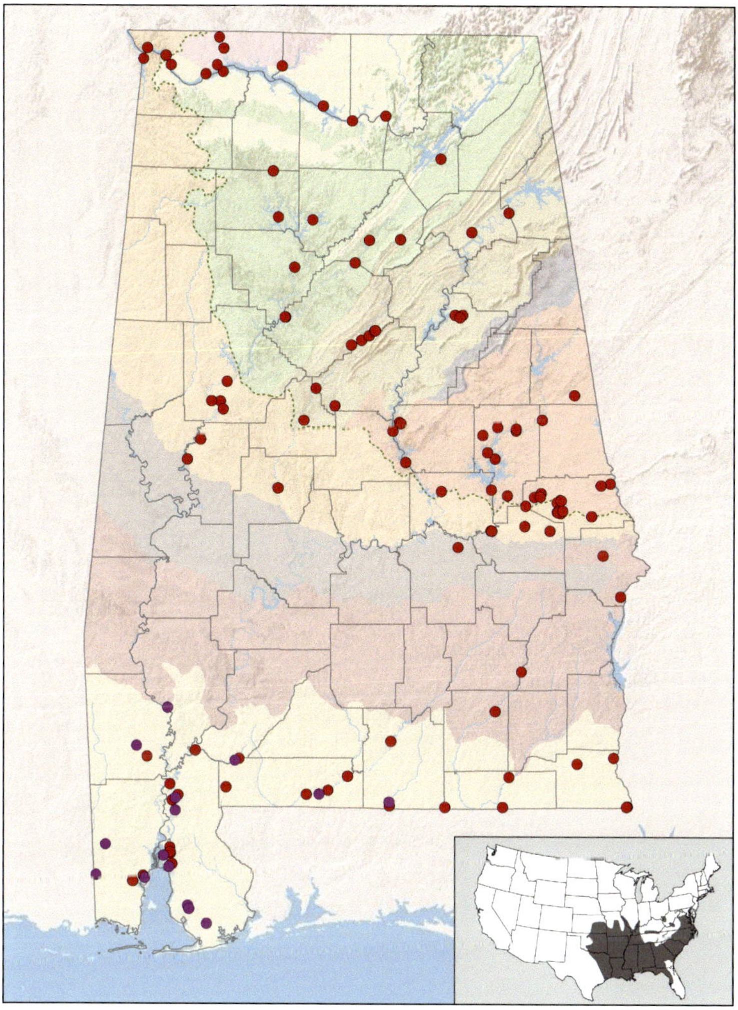

Alabama and continental US distribution of the River Cooter; purple dots indicate hybrids with the Florida Cooter

have testes that reach peak size in the fall so that sperm are stored over winter in the epididymides; when males emerge in spring, they have viable sperm but testes that are at their smallest size. Females begin development of ovarian follicles in early July, and then these follicles enlarge through fall, maintain size over winter, and reach peak size and ovulate in late May to early June of the following year. In early spring, males court females by approaching them in water while vibrating their greatly elongated front toenails against a female's head. Receptive females likely select from among potential mates, generally in spring, but with scattered matings occurring throughout the season of activity. Nesting occurs in sandy, friable soil on elevated sites near the water in some females and in a variety of habitats far from water in others. Each female lays two clutches per year with as many as nineteen eggs per clutch. Most eggs hatch in fall and remain in the nest overwinter, emerging the following spring. However, some emerge in fall, generally if the nest is disturbed.

Males mature at five years of age when they are about 6.5 inches (160 mm) in carapace length; females mature at age fourteen when they are about 5.5 inches (300 mm) in carapace length. Population density can reach such high levels that all exposed rocks and logs in some stretches of major rivers in Alabama will be covered with this species. River Cooters hibernate during the coldest months of the year in most parts of Alabama, but in some streams in the Lower Coastal Plain they are active virtually year round.

Adult River Cooters are almost entirely herbivorous, feeding on a wide variety of plant material. But Riverweed (*Podostemon ceratophyllum*), when it is available, dominates the diet. Juvenile River Cooters are omnivorous.

Conservation and Management River Cooters can be abundant in free-flowing waters and are widely distributed throughout the state of Alabama. But because all turtle species can suffer from overharvest, it is unlawful to capture River Cooters for sale. An individual may possess up to two River Cooters per day for personal use only. Some aspects of urbanization may increase the abundance of River Cooters in urban streams (Barrett and Guyer 2008). Because a significant portion of females nest far from the river's edge, current best management practices for streamside management zones (a fifty-foot buffer zone along streams; Alabama Forestry Commission 2007) are likely to be inadequate to protect nesting habitat for this species. For managers wishing to minimize the effects of management practices on River Cooters, any use of heavy equipment in the streamside management zone should be done in April and May (after hatchling emergence but before most nests are initiated). Basking sites can be limiting, so efforts to supplement these are likely to be beneficial to populations. River Cooters are eaten by some Alabamians, especially in the southwestern portion of the state. However, this source of mortality apparently is not strong enough to adversely affect populations of this species.

Taxonomy Mount's (1975) examination of this species in Alabama suggested that it is widespread with five subspecies. However, two of these subspecies have been elevated to species status, *P. suwanniensis* and *P. texana*. To add further taxonomic confusion to this group, one previously accepted species, *P. floridana*, was determined to be genetically indistinguishable from *P. concinna*, leading to an attempt to synonymize these two taxa (Seidel 1994). Accumulation of DNA sequence data from the nuclear genome has continued to paint a confusing picture of the relationships of *Pseudemys* (Spinks et al. 2009). Because forms showing the color pattern and shell morphology of *P. floridana* occupy a contiguous geographic area and are found in more lentic waters than *P. concinna*, we retain this as a separate species that is sister to *P. concinna*. Hybridization between *P. concinna* and *P. floridana* occurs in Alabama, most frequently in the extreme southwestern portion of the state. The populations inhabiting most of the small streams entering Mobile Bay also include many hybrid individuals. We accept the conclusion of Mount (1975) that *P. c. concinna*, *P. c. hieroglyphica*, and *P. c. mobilensis* in Alabama are insufficiently distinctive to warrant subspecific status. Thus, we consider *P. concinna* to be one species with no recognizable subspecies (see also Fritz and Havas 2007). Previous authors have placed this species in the genera *Chrysemys*, *Emys*, *Ptychemys*, and *Testudo*.

Florida Cooter

Pseudemys floridana (Le Conte, 1830)

Description Florida Cooters are large, robust aquatic turtles attaining a maximum carapace length of about 15.5 inches (400 mm) in females and 11 inches (275 mm) in males. The carapace is oval, relatively high-domed in females, and somewhat serrate and often flaring along its posterior border. In color, the carapace is dark brown to nearly black with yellowish markings, the most conspicuous of which is a vertical bar, on the second pleural scute, that is forked at either or both ends (*P. concinna* usually has a C-shaped light mark on the second pleural). The plastron is immaculate yellow, and the bridge is plain or has one or more dark spots. On their ventral side, the anterior marginals have dark spots that may or may not have light centers; the undersurfaces of the posterior marginals are unmarked (in contrast to other species of *Pseudemys* in Alabama). The head, limbs, and tail are black, or nearly so, with yellow stripes. A transverse dark bar occurs across the iris. In hatchlings, the carapace is greenish above and has numerous yellow markings, but the vertical mark on the second pleural scute is not as bold as in adults.

Alabama Distribution This species is found across the Lower Coastal Plain. Specimens from the eastern portion of this range lack evidence

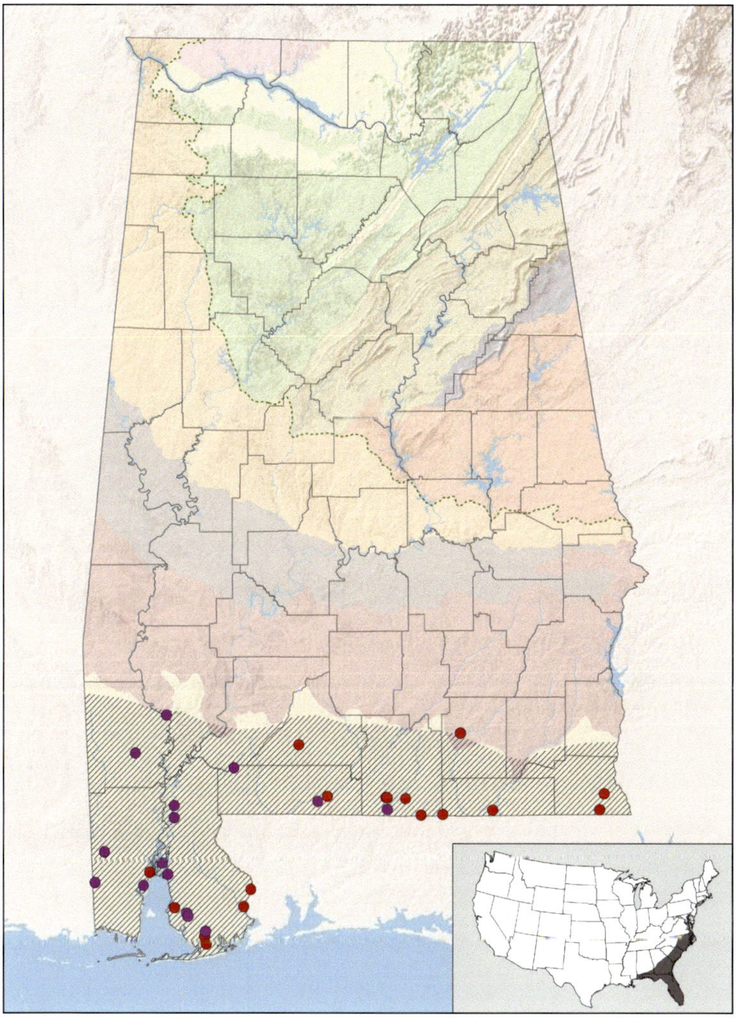

Left: Alabama and continental US distribution of the Florida Cooter; purple dots indicate hybrids with the River Cooter

Opposite: Adult female Florida Cooter, Covington County, AL

of hybridization, but specimens from around Mobile Bay and western Mobile County show high proportions of apparent hybrids.

Habits In Alabama *P. floridana* occurs in greatest numbers in large ponds and lakes with abundant aquatic vegetation. Lake Jackson, a large sinkhole lake in Covington County, provides particularly favorable habitat. This turtle also occurs in oxbows and sloughs associated with rivers and in some of the small, sluggish rivers, such as the Perdido, that have aquatic vegetation along their edges. It is seldom seen in the main channels of relatively deep, swift Coastal Plain streams, such as the Conecuh and Choctawhatchee Rivers, a habitat type well suited to *P. concinna.* Nesting occurs from late May until August (Thomas 1972). Clutch size ranges from four to twenty-two, with most females laying two or more times each season. Because of its similarity to *P. concinna,* this pattern of reproduction likely generates some clutches that hatch in the fall and other clutches that overwinter in the nest and emerge the following spring. The latter likely predominates since hatchlings are seen in greatest abundance in the spring. Sandy areas are chosen as nesting sites. Nest chambers for this species in Alabama usually consist of a single hole, but some females dig three holes, one in the center and one on each side (Aresco 2004; Thomas 1972). During the first year, the young turtles feed on both plant and animal life. After that they become almost totally herbivorous, feeding on a wide variety of plants.

Conservation and Management Florida Cooters can be abundant. Nevertheless, because populations also are sensitive to overharvest, it is illegal to capture this species for sale. Two individuals may be possessed per day for personal use. Current best management practices for streamside management zones are likely to be adequate for this species, especially if sand bars along slow-moving rivers are protected from off-road-vehicle traffic. For managers wishing to minimize the effects of management on Florida Cooters, any use of heavy equipment in the streamside management zone should be done in April and May to avoid damage to nests. Basking sites can be limiting, so efforts to supplement these are likely to benefit populations of Florida Cooters. Finally, because members of this species, especially the females, cross roads frequently (Steen et al. 2006), wildlife fencing and designed underpasses in areas of heavy traffic may be necessary to maintain the species in urban settings.

Taxonomy Accumulating sequence data from the nuclear genome fail to distinguish *P. floridana* from *P. concinna* (Spinks et al. 2009). However, because forms showing the color pattern and shell morphology of *P. floridana* occupy a contiguous geographic area and are found in more lentic waters than *P. concinna,* we retain *P. floridana* as a separate species and consider it to be the sister taxon to *P. concinna.* Hybridization between *P. concinna* and *P. floridana* occurs in Alabama, most frequently in the extreme southwestern portion of the state. The populations inhabiting most of the small streams entering Mobile Bay contain some members that are hybrids. Three subspecies of *P. floridana* have been recognized. Two of these, *P. f. floridana* and *P. f. hoyi*, have been reported to occur in Alabama, but no substantial evidence exists of the latter's presence in the state and we do not consider it to be a viable taxon. The third subspecies, *P. f. pennisularis*, has been elevated to species status in recent studies (e.g., Spinks et al. 2009). Therefore, we consider *P. floridana* to be a species with no subspecific variation. Previous authors have placed this species in the genera *Chrysemys*, *Emys*, and *Testudo*.

Appendix 1

List of subspecies from Mount (1975) now recognized as full species.

TAXON	SCIENTIFIC NAME	SOURCE
Emydidae	*Chrysemys picta dorsalis*	Starkey et al. 2003
	Graptemys pseudogeographica ouachitensis	Vogt 1993
Kinosternidae	*Sternotherus minor depressus*	Walker 1998a
	Sternotherus minor peltifer	Ernst et al. 1998
	Sternotherus minor minor	Ernst et al. 1998
Trionychidae	*Apalone mutica calvata*	This volume

Appendix 2

List of turtle species distributed within (or nearly so) the Coastal Plain of Alabama.

FAMILY	SPECIES
Cheloniidae	*Caretta caretta*
	Chelonia mydas
	Eretmochelys imbricata imbricata
	Lepidochelys kempii
Dermochelyidae	*Dermochelys coriaceae*
Emydidae	*Chrysemys dorsalis*
	Deirochelys reticularia reticularia
	Graptemys barbouri
	Graptemys ernsti
	Graptemys nigrinoda delticola
	Graptemys nigrinoda nigrinoda
	Graptemys pulchra
	Malaclemys terrapin pileata
	Pseudemys alabamensis
	Pseudemys floridana
	Terrapene bauri
	Trachemys scripta scripta

FAMILY	SPECIES
Kinosternidae	*Kinosternon baurii*
	Kinosternon subrubrum hippocrepis
	Sternotherus carinatus
	Sternotherus minor
Testudinidae	*Gopherus polyphemus*
Trionychidae	*Apalone calvata*
	Apalone ferox

Glossary

Amniotes: a group of vertebrates that includes mammals and reptiles (including birds). This group is characterized by the presence of a fertilized egg that develops an amnion, chorion, and allantois, layers that serve to bathe and protect a developing embryo as well as exchange gases and store waste products as the embryo grows.

Amphibian: a group of land vertebrates with at most four fingers and five toes, smooth moist skin, and eggs lacking the amnion, chorion, and allantois of an amniote. This group contains many fossil radiations and all of the living frogs, salamanders, and caecilians.

Anastomosing: an irregular, web-like pattern of branching and reconnecting colors or segments.

Anterior: toward the front or head end of an animal.

Anterolateral: toward the front but along the sides of an animal.

Anthozoan: a major group of largely marine animals called sea anemones, corals, and sea pens. These animals have a tubular shape with a series of tentacles radiating from the end that forms the mouth.

Aquatic: associated with water.

Arthropod: any member of the group Arthropoda, a large radiation of animals with jointed legs.

Barbels: fleshy, fingerlike projections found on the chin of some turtles.

Bifurcation: a pattern in which a single branch divides into two branches.

Binomial: two names; referring to a system in which each species is identified by a genus name (a group of closely related species to which a particular species belongs) and a specific epithet (a name particular to a member of a genus).

Biogeography: the scientific study of the patterns and processes by which living and fossil organisms have become distributed the way that they are.

Biota: all living organisms of a particular place.

Blotch: a large, dark color marking with irregular rounded edges.

Caecum: an outpocketing of the digestive system where the small intestine joins the large intestine.

Calcareous: a structure hardened by deposition of calcium.

Carapace: the dorsal (top) half of a turtle shell.

Carnivorous: a foraging mode in which animals are the primary diet item.

Carrion: flesh of dead vertebrates.

Clade: a monophyletic group of organisms

Cloaca: the single opening that transmits digestive and urinary products to the outside of the body of a vertebrate.

Clutch: all the eggs laid by a female in a single nesting event.

Cnidarian: any member of the group Cnidaria, the large radiation of animals that includes the jellyfishes, sea anemones, and hydras.

Conservation: the field of biological sciences focused on retaining rare organisms and their communities.

Crustacean: an organism belonging to the group Crustacea: crabs, crayfish, lobsters, shrimp, isopods, water fleas, and barnacles.

Cryptodiran: a group of turtles capable of withdrawing the head and neck into the shell via formation of an S-shape to the neck vertebrae.

Cusp: a pointed, toothlike projection.

Dendritic: markings that branch in a treelike fashion.

Diapause: a metabolic state of dormancy achieved by some animals during periods of environmental stress.

Dichotomous: a choice involving only two possibilities.

Dichromatism: occurrence of two distinct color patterns within a species, typically associated with differing colors between sexes.

Dimorphism: occurrence of two distinct sizes or shapes of organisms within a species, typically associated with differences between sexes.

Dorsoventrally: along the axis of the body separating the dorsal from the ventral half.

Dorsum: the top or back of an animal.

Echinoderm: any member of the group Echinodermata, the radiation of organisms that includes the starfishes, sea urchins, and sea cucumbers.

Ecosystem: a community of organisms and the non-living environmental features that interconnect the organisms.

Emydid: member of the turtle family Emydidae.

Estuary: a coastal wetland area formed by the transition of the freshwater of a river to the saltwater of an ocean.

Extant: an organism that survives in the modern fauna.

Fall line: a geological formation characterized by relatively steep rocky relief where the relatively flat Coastal Plains meet the more mountainous terrain of the Appalachian Mountains.

Family: a group of closely related genera.

Fauna: a term for all animals of a particular area.

Foraging: the process of seeking and consuming food by animals.

Forelobe: a projection oriented toward the anterior end of an animal.

Gene flow: the process by which genetic material is transferred among individuals within a population due to sexual reproduction.

Genetic clade: a lineage of organisms identifiable by unique genetic character states; cryptic species.

Genome: the genetic material of an organism.

Genus: a group of closely related species.

Gravid: the condition of having shelled eggs in the uterine tract of a female.

Gular: the chin region of a vertebrate.

Haplotype: a combination of adjacent alleles that are inherited as a unit.

Hatchling: the stage of development of an animal immediately after it hatches from an egg.

Herbivorous: feeding on plants.

Herpetofaunal: pertaining to all amphibian and reptile species of a particular area.

Herpetology: the study of amphibians and reptiles.

Hybrid: an organism possessing the genome of two or more species.

Incubation: for turtles, the time period between deposition of eggs and hatching of those eggs.

Indigenous: native to a particular area.

Insectivorous: organisms that prey principally on insects.

Interorbital: the region between the eyes.

Invertebrate: any animal that does not belong to the group Vertebrata; animals lacking a vertebral column.

Juvenile: an organism that is no longer a hatchling but is not yet sexually mature.

Keel: a long ridge along an anatomical feature; in turtles, any ridge running the length of the carapace.

Laminae: broad, flat sheets of tissue.

Lentic: related to or living in still waters.

Lineage: a group of organisms that share a common origin in evolutionary history, exist over time, and eventually cease to exist due to extinction.

Linnean: referring to the classification produced by Linnaeus.

Longitudinal: oriented along the long axis of the body.

Mammal: a member of the group Mammalia.

Marginals: epidermal scutes of turtles located along the outer edge of the shell.

Marsh: a flat, open, wetland dominated by grasses and sedges.

Mesozoic: a time period of the geological record from 66 to 250 million years before the present.

Mid-dorsal: located along the middle of the back.

Migration: a long-distance movement taking an organism or group of organisms from one home range to another.

Mitochondria: organelles within the cell of vertebrates that produces energy-generating molecules.

Mollusk: any member of the group Mollusca, the snails, slugs, and octopi.

Monophyletic: a group containing an ancestral species and all descendants of that ancestor.

Morphology: the shape or anatomical outline of an organism.

Morphometric: a series of measures of the body that are used to determine size and shape.

Nares: openings leading into the nasal cavities.

Nictitating membrane: an opaque structure similar in morphology and function to eyelids.

Nominate: a subspecific name that repeats the species name.

Nuchal: an epidermal scute of the carapace located above the base of the neck.

Nuclear DNA: genetic material found in the nucleus of a cell.

Ocelli: color pattern of light and dark, eyelike rings.

Offspring: the organisms produced by a female via reproduction.

Omnivorous: feeding on a variety of plant and animal material.

Organism: any living individual.

Overstory: that portion of the vegetation associated with the canopy created by trees and tall shrubs.

Oviparous: a reproductive mode in which fertilized eggs are placed in a nest.

Ovulation: the process by which oocytes (eggs) develop and are expelled from the ovary into the uterine tract as ova.

Oxbow: a ponded wetland created by the meandering of rivers in flat floodplains.

Paleocene: a period in the geological record from 56 to 66 million years before the present.

Papillae: small, fingerlike projections of skin.

Paramedial: color markings located to either side of a line along the mid-dorsal axis of the body.

Phylogenetic: referring to the ancestor-descendant relationships that exist among organisms.

Physiographic: biotic regions characterized by distinct features of the soils and topography.

Plastron/Plastral: the portion of the shell of a turtle that covers the belly.

Pleural: a row of scutes on the carapace located between the vertebrals and marginals.

Pleurodira: a group of turtles characterized by a specialized neck joint that allows them to position the head and neck along the side of the body. These are referred to as side-necked turtles.

Polytypic: a species that consists of distinct geographic color morphs.

Posterior: toward or at the tail end of a vertebrate.

Postorbital: a position just behind the eye.

Predator: an animal that kills and consumes other animals.

Proventriculus: a muscular area of the esophagus immediately anterior to the stomach; the gizzard.

Pygal: a scute of the carapace located just above the base of the tail.

Radiation: a group of organisms diverging from a common ancestor.

Reproduction: the process by which an individual organism creates other individual organisms. In turtles, reproduction is sexual, involving passage of sperm from a male to the reproductive tract of a female, where it fertilizes an ovum produced by the female.

Reptile: the ancestral amniote possessing epidermal scales of alpha and beta keratin, accommodation of the lens via striated muscles, and nasal glands located outside the nasal cavity, and all descendants of that ancestor. This group contains the living turtles, crocodilians, squamates, and birds of the world.

Riparian zone: the band of vegetation along a river or stream. Typically, this zone is forested with trees that require constant access to water.

Sandbar: a region of a stream, typically on the downstream side of a bend, where sand accumulates. Sandbars create nesting habitat for many turtle species.

Scale: a thickened and folded outermost portion of the skin of many vertebrates. In turtles, a scale is composed of keratin, a protein that makes the scale impervious to water.

Scute: a large, platelike scale in reptiles; in turtles, forming the outermost covering for shells.

Septal: of the skin that separates the right and left nasal cavities.

Serrate: an edge with numerous pointed, sawlike projections.

Sex: a category containing two states, male and female, identifying the role played by an individual during sexual reproduction.

Slough: an area of sluggish, muddy water associated with backwater of slow-moving rivers.

Species: the fundamental unit of biological organization containing individuals that are more closely related to each other than they are to other species and that are characterized by diagnostic features indicating their unique origin in time.

Sperm: gametes produced by male organisms.

Subdentary: the region just below the lower jaw.

Suborder: a category of the Linnean hierarchy that contains a group of closely related families, but not all families, within an order.

Subspecies: a geographic variant within a species, characterized by unique color patterns or morphological features, but grading into other variants because of interbreeding between adjacent subspecies.

Substrate: a general term for the surfaces upon which an organism may rest.

Superfamily: a category of the Linnean hierarchy that consists of a group of closely related families, but not all families within an order or suborder.

Supramarginal: a row of scutes located on the carapace between the marginal and pleural scutes.

Suture: a joint where two anatomical features meet, as in the joint between bones or the joint between scutes of a turtle shell.

Systematics: the study of evolutionary interrelationships of species.

Taxa/Taxon: a group of closely related organisms; the group might be as restricted as individuals within a single species or as inclusive as species within an entire Kingdom.

Taxonomy: the science associated with discovering new species and naming them.

Tectonic: geological activities causing the movement of the earth's plates.

Telemetry: the study of movements of organisms by use of radio transmitters attached to them, allowing them to be relocated.

Terminus: the end of an object.

Terrapin: a member of the genus *Malaclemys* (scientific meaning), or any terrestrial turtle (vernacular meaning).

Terrestrial: organisms that are associated with land.

Tetrapod: any member of the group Tetrapoda, those vertebrates characterized by the presence of land limbs or being derived from an ancestor with limbs.

Topography: the three-dimensional surface of the earth.

Tributary: a branch of a major river or stream.

Tubercle: a fleshy bump.

Tuberculate: a surface covered with tubercles.

Tubular proboscis: a fleshy, elongate projection creating the nose of softshell turtles.

Vascularized: characterized by the presence of many blood vessels, especially capillaries.

Ventral: toward the bottom or belly of an animal.

Vermiculation: worm-like color marking.

Vertebral: associated with or toward the vertebrae (backbones).

Vertebrate: a group of animals characterized by the presence of a series of bony elements creating a vertebral column (backbone). This group includes lampreys, fishes, amphibians, mammals, and reptiles (including birds), among living forms.

Wetland: an area characterized by plants that require wet soils. These areas typically have visible water present, but also include sites, like bogs, where water may be below the soil surface.

Notes on Illustrations

Photographs Courtesy of:

Mark A. Bailey, pages 85, 86, 87, 126
Roger D. Birkhead, page 168
Alan Cressler, cover and pages 10, 76, 97, 176, 210
Helen Czech, page 123
Brian P. Folt, page 13
Jason Folt, pages 53, 114
James C. Godwin, pages 71, 79, 100, 117, 133, 148, 149, 183, 184, 187, 188, 196, 197, 200, 201, 209, 216, 217, 221, 222
Scott Gravette, page 165
Craig Guyer, pages 11, 12
Pierson Hill, pages 65, 109, 152, 179
Robert H. Mount, page 192
Joseph B. Pfaller, pages 33, 34, 40, 41, 47, 48, 57, 58
Lisa Powers, page 124
Corey Raimond, page 68
Greg Sievert, page 156
Eric Soehren, pages 96, 137, 161
James A. Stiles, page 226
Kenneth Wray, pages 80, 129, 175, 180, 206

Drawings Courtesy of Claire C. Floyd

page 29: bottom right, based on AUM 17078
page 30: bottom right, based on AUM 2836
page 38: bottom left, based on AUM 28876
page 63: top left, based on AUM 12443; top right, based on AUM 4980; bottom left, based on AUM 11802; bottom right, based on AUM 12433
page 64: top, based on AUM 4980
page 64: bottom, based on AUM 8951
page 75: left, based on AUM 1984; right, based on AUM 8951
page 106: left, based on AUM 39648; right, based on AUM 8936
page 108: left, based on AUM 39648; right, based on AUM 8988
page 113: top, based on AUM 8987
page 113: bottom, based on AUM 7790
page 121: top left, based on AUM 8935; top right, based on AUM 6277; bottom left, based on AUM 9537; bottom right, based on AUM 9537

page 122: top left, based on AUM 13744; top right, based on AUM 6277; bottom left, based on AUM 13744; bottom right, based on AUM 17076
page 142: top right and bottom left, based on AUM 6135; top left, based on AUM 2836
page 143: top left, based on AUM 7577
page 144: top right and bottom right, based on AUM 17139
page 145: top left, based on AUM 25147; top right, based on AUM 8986; bottom left, based on AUM 7577; bottom right, based on AUM 7577
page 173: top left, based on AUM 5977
page 174: top right, based on AUM 5008
page 205: left, based on AUM 25147; right, based on AUM 8986
page 214: bottom left, based on AUM 8986; bottom right, based on AUM 7477
page 215: left, based on AUM 7577

Maps by Mark A. Bailey and Craig Guyer

page 35: Solid dots indicate records believed by authors to be valid.
pages 42, 49, 59: Solid dots indicate museum specimens examined by authors or records from the Share the Beach program believed by the authors to be valid (see species account).
page 54: Solid dot indicates nearest sighting believed by authors to be valid.
pages 66, 69, 72, 77, 81, 98, 101, 115, 118, 125, 127, 150, 153, 157, 162, 177, 185, 218: Solid dots indicate localities of specimens or photographs examined by authors or ADCNR/Natural Heritage Program occurrence records believed to be valid.
page 88: Solid dots indicate (1) localities of specimens or photographs examined by authors, (2) ADCNR/Natural Heritage Program occurrence records believed to be valid, and (3) localities of Gopher Tortoise burrows examined by the authors.
page 110: Solid dot indicates locality of specimen examined by authors.
page 130: Red dots indicate records of museum specimens or photographs examined by authors or ADCNR/Natural Heritage Program occurrence records believed to be valid and referable to this species. Purple dots indicate specimens displaying genetic evidence of hybridization with *Sternotherus peltifer*, the Stripe-necked Musk Turtle.
page 134: Red dots indicate records of museum specimens or photographs examined by authors or ADCNR/Natural Heritage Program occurrence records believed to be valid and referable to this species. Purple dots indicate specimens displaying alternate color patterns, perhaps indicative of intraspecific variation.
page 138: Red dots indicate localities of specimens or photographs examined by authors or ADCNR/Natural Heritage Program occurrence records believed to be valid. Purple dots indicate specimens displaying genetic

evidence of hybridization with *Sternotherus depressus*, the Flattened Musk Turtle.

page 166: Red dots indicate records of museum specimens or photographs examined by authors or ADCNR/Natural Heritage Program occurrence records believed to be valid and referable to this species. Dots with a black center indicate specimens displaying features of *Chrysemys picta marginata*, the Midland Painted Turtle. Based on genetic evidence from Starkey et al. (2003) we consider these specimens to be *C. dorsalis*, but with morphological evidence of past gene flow with *C. p. picta*.

page 169: Red dots indicate records of museum specimens or photographs examined by authors or ADCNR/Natural Heritage Program occurrence records believed to be valid and referable to this species. Purple dots indicate specimens displaying evidence of hybridization with *Chrysemys dorsalis*, the Southern Painted Turtle.

page 181: Red dots indicate records of museum specimens or photographs examined by authors or ADCNR/Natural Heritage Program occurrence records believed to be valid and referable to this species. Purple dots indicate specimens displaying evidence of hybridization with *Graptemys ernsti*, the Escambia Map Turtle.

page 189: Red dots indicate records of museum specimens or photographs examined by authors or ADCNR/Natural Heritage Program occurrence records believed to be valid and referable to this species. Purple dots indicate specimens displaying evidence of hybridization with *Graptemys barbouri*, Barbour's Map Turtle.

page 193: Red dots indicate localities of specimens or photographs examined by authors or ADCNR/Natural Heritage Program occurrence records believed to be valid. Red dot with black center indicates a valid record likely representing a waif dispersing via Tennessee-Tombigbee waterway.

page 198: Red dots indicate records of museum specimens or photographs examined by authors or ADCNR/Natural Heritage Program occurrence records believed to be valid and referable to this subspecies. Purple dots indicate specimens displaying evidence of intergradation with *Graptemys nigrinoda delticola*, the Southern Black-knobbed Sawback.

page 202: Red dots indicate records of museum specimens or photographs examined by authors or ADCNR/Natural Heritage Program occurrence records believed to be valid and referable to this subspecies. Purple dots indicate specimens displaying evidence of intergradation with *Graptemys nigrinoda nigrinoda*, the Northern Black-knobbed Sawback.

page 207: Red dots indicate records of museum specimens or photographs examined by authors or ADCNR/Natural Heritage Program occurrence records believed to be valid and referable to this subspecies. Purple dots indicate specimens displaying evidence of intergradation with *Trachemys scripta scripta,* the Yellow-bellied Slider.

page 211: Red dots indicate records of museum specimens or photographs examined by authors or ADCNR/Natural Heritage Program occurrence records believed to be valid and referable to this subspecies. Purple dots indicate specimens displaying evidence of intergradation with *Trachemys scripta elegans,* the Red-eared Slider.

page 223: Red dots indicate localities of specimens or photographs examined by authors or ADCNR/Natural Heritage Program occurrence records believed to be valid. Purple dots indicate specimens displaying evidence of hybridization with *Pseudemys floridana,* the Florida Cooter.

page 227: Red dots indicate records of museum specimens or photographs examined by authors or ADCNR/Natural Heritage Program occurrence records believed to be valid and referable to this species. Purple dots indicate specimens displaying evidence of hybridization with *Pseudemys concinna,* the River Cooter.

References

Alabama Department of Conservation and Natural Resources, Division of Wildlife and Freshwater Fisheries. 2005. "Conserving Alabama's Wildlife: a Comprehensive Strategy." Alabama Department of Conservation and Natural Resources, Montgomery.

Alabama Forestry Commission. 2007. "Alabama's Best Management Practices for Forestry." http://www.forestry.alabama.gov/publications/bmps/2007_bmp_manual.pdf.

Arata, A. A. 1958. "Notes on the Eggs and Young of *Gopherus polyphemus* (Daudin)." *Quarterly Journal of the Florida Academy of Sciences* 21: 274–280.

Aresco, M. J. 2004. "Reproductive Ecology of *Pseudemys floridana* and *Trachemys scripta* Testudines: Emydidae) in Northwestern Florida." *Journal of Herpetology* 38: 249–256.

Aresco, M. J., and J. L. Dobie. 2000. "Variation in Shell Arching and Sexual Size Dimorphism of River Cooter, *Pseudemys concinna*, from Two River Systems in Alabama." *Journal of Herpetology* 34: 313–317.

Ashton, K. G., and C. R. Feldman. 2003. "Bergmann's Rule in Non-Avian Reptiles: Turtles Follow It, Lizards and Snakes Reverse It." *Evolution* 57: 1151–1163.

Auffenburg, W. G. 1966. "On the Courtship of *Gopherus polyphemus*." *Herpetologica* 22: 113–117.

Bailey, K. A., and M. A. Bailey. 2002. "Nesting Study of the Flattened Musk Turtle, *Sternotherus depressus*." Final Report, Grant Agreement no. 1448–40181–00-G-118, US Fish and Wildlife Service.

Bailey, K. A., and C. Guyer. 1998. Demography and Population Status of the Flattened Musk Turtle, *Sternotherus depressus*, in the Black Warrior Basin of Alabama." *Chelonian Conservation and Biology* 3: 77–83.

Barrett, R. K., and C. Guyer. 2008. "Differential Responses of Amphibians and Reptiles in Riparian and Stream Habitats to Land Use Disturbances in Western Georgia, USA." *Biological Conservation* 141: 2290–2300.

Birkhead, R. D., C. Guyer, S. M. Hermann, and W. K. Michener. 2005. "Patterns of Folivory and Seed Ingestion by Gopher Tortoises (*Gopherus polyphemus*)." *American Midland Naturalist* 154: 143–151.

Boglioli, M. D., W. K. Michener, and C. Guyer. 2000. "Habitat Selection and Modification by the Gopher Tortoise, *Gopherus polyphemus*." *Chelonian Conservation and Biology* 3: 699–705.

Bollmer, J. L., M. E. Irwin, J. P. Rieder, and P. G. Parker. 1999. "Multiple Paternity in Loggerhead Turtle Clutches." *Copeia* 1999: 475–478.

Bowen, B. W., and S. A. Karl. 2007. "Population Genetics and Phylogeography of Sea Turtles." *Molecular Ecology* 16: 4886–4907.

Buhlmann, K. A. 1995. "Habitat Use, Terrestrial Movements, and Conservation of the Turtle, *Deirochelys reticularia* in Virginia." *Journal of Herpetology* 29: 173–181.

Buhlmann, K. A., T. K. Lynch, J. W. Gibbons, and J. L. Greene. 1995. "Prolonged Egg Retention in the Turtle *Deirochelys reticularia* in South Carolina." *Herpetologica* 51: 457–462.

Buhlmann, K. A., J. W. Gibbons, and D. R. Jackson. 2008. "*Deirochelys reticularia* (Latrielle 1801)—Chicken Turtle." *Chelonian Research Monographs* 5:Doi:10.3854/Crm.5.014.Reticularis.V1.2008.

Burke, R. L., M. A. Ewert, J. B. McLemore, and D. R. Jackson. 1996. "Temperature-Dependent Sex Determination and Hatching Success in the Gopher Tortoise (*Gopherus polyphemus*)." *Chelonian Conservation and Biology* 2: 86–88.

Butler, J. M., C. K. Dodd, M. Aresco, and J. D. Austin. 2011. "Morphological and Molecular Evidence Indicates That the Gulf Coast Box Turtle (*Terrapene carolina major*) Is Not a Distinct Evolutionary Lineage in the Florida Panhandle." *Biological Journal of the Linnean Society* 102: 889–901.

Carr, A. F. 1952. "Handbook of Turtles: The Turtles of the United States, Canada, and Baja California." Ithaca, NY: Comstock Publishing Associates. 542 pp.

Carr, A. F., and S. Stancyk. 1975. "Observations on the Ecology and Survival Outlook of the Hawksbill Turtle." *Biological Conservation* 8: 161–172.

Close, D. K. 1982. "The Reproductive Cycle of *Sternotherus minor depressus*." Master's thesis, University of Alabama, Birmingham. 101 pp.

Congdon, J. D., J. W. Gibbons, and J. L. Greene. 1983. "Parental Investment in the Chicken Turtle (*Deirochelys reticularia*)." *Ecology* 64: 419–425.

Congdon, J. D., G. L. Breitenbach, R. C. Van Loben Sels, and D. W. Tinkle. 1987. "Reproduction and Nesting Ecology of Snapping Turtles *Chelydra serpentina* in Southeastern Michigan." *Herpetologica* 43: 39–54.

Cox, W. A., and K. R. Marion. 1978. "Observations on the Female Reproductive Cycle and Associated Phenomena in Spring-Dwelling Populations of *Sternotherus minor* in North Florida (Reptilia: Testudines)." Herpetologica 34:20-33.

Cox, W. A., J. B. Hazelrig, M. E. Turner, R. A. Angus, and K. R. Marion. 1991. "A Model for Growth in the Musk Turtle, *Sternotherus minor*, in a North Florida Spring." *Copeia* 1991: 954–968.

Crother, B. I. (ed.). 2012. "Scientific and Standard English Common Names of Amphibians and Reptiles of North America North of Mexico, with Comments Regarding Confidence in Our Understanding." *SSAR Herpetological Circular* no. 39: 1–92.

Davidson, J. W. 1971. "Geographic Variation in the Pond Slider, *Pseudemys scripta*, in Alabama." Master's thesis, Auburn University. 56 pp.

De Queiroz, K. 1998. "A General Lineage Concept of Species, Species Criteria, and the Process of Speciation: a Conceptual Unification and Terminological Recommendations." In *Endless Forms: Species and Speciation*, ed. D. J. Howard and S. H. Berlocher, 57–75. Oxford: Oxford University Press.

De Queiroz, K., and J. Gautier. 1992. "Phylogenetic Taxonomy." *Annual Review of Ecology and Systematics* 23: 449–480.

Dodd, C. K. 1990. "Effects of Habitat Fragmentation on a Stream-Dwelling Species, the Flattened Musk Turtle *Sternotherus depressus*." *Biological Conservation* 54: 33–45.

Dodd, C. K., and G. S. Morgan. 1992. "Fossil Sea Turtles from the Early Pliocene Bone Valley Formation, Central Florida." *Journal of Herpetology* 26: 1–8.

Dodd, C. K., K. M. Enge, and J. N. Stuart. 1988. "Aspects of the Biology of the Flattened Musk Turtle, *Sternotherus depressus*, in Northern Alabama." *Bulletin of the Florida State Museum Biological Science* 34: 1–64.

Doody, J. S. 1995. "A Comparative Nesting Study of Two Syntopic Species of Softshell Turtles (*Apalone mutica and Apalone spinifera*) in Southcentral Louisiana." Master's thesis, Southeastern Louisiana University. 134 pp.

Dutton, P. H., B. W. Bowen, D. W. Owens, A. Baragan, and S. K. Davis. 1999. "Global Biogeography of the Leatherback Turtle (*Dermochelys coriacea*)." *Journal of Zoology* 248: 397–409.

Echelle, A. A., J. C. Hackler, J. B. Lack, S. R. Ballard, J. Roman, S. F. Fox, D. M. Leslie Jr., and R. A. Van Den Bussche. 2010. "Conservation Genetics of the Alligator Snapping Turtle: Cytonuclear Evidence of Range-Wide Bottleneck Effects and Unusually Pronounced Geographic Structure." *Conservation Genetics* 11: 1375–1387.

Elliot, M. J., J. W. Gibbons, C. D. Camp, and J. B. Jensen. 2008. *Amphibians and Reptiles of Georgia*. Athens: University of Georgia Press.

Engstrom, T. N., S. B. Shaffer, and W. P. McCord. 2004. "Multiple Datasets, High Homoplasy, and the Phylogeny of Softshell Turtles (Testudines: Trionychidae)." *Systematic Biology* 53: 693–710.

Ennen, J. R., B. R. Kreiser, and C. P. Qualls. 2010. "Low Genetic Diversity in Several Gopher Tortoise (*Gopherus polyphemus*) Populations in the Desoto National Forest, Mississippi." *Herpetologica* 66: 31–38.

Ennen, J. R., M. E. Kalis, A. L. Patterson, B. R. Kreiser, J. E. Lovich, J. Godwin, and C. P. Qualls. 2014. "Clinal variation or validation of a subspecies? A case study of the *Graptemys nigrinoda* complex (Testudines: Emydidae)." Biological Journal of the Linnean Society 111:810-822.

Ernst, C. H., and J. E. Lovich. 2009. *Turtles of the United States and Canada*. Baltimore: Johns Hopkins University Press.

Ernst, C. H., L. J. Miller, K. R. Marion, and W. A. Cox. 1988. "Comparisons of Shell Morphology among Turtles of the *Kinosternon minor* Complex." *American Midland Naturalist* 120: 282–288.

Estridge, R. E. 1970. "The Taxonomic Status of *Sternothaerus depressus* (Testudinata, Kinosternidae) with Observations on Its Ecology." Master's thesis, Auburn University. 49 pp.

Fahey, K. M. 1987. "Aspects of the Life History of the River Cooter, *Pseudemys concinna* (Le Conte), in the Tallapoosa River, Tallapoosa County, Alabama." PhD diss., Auburn University. 155 pp.

Folkerts, G. W., and R. H. Mount. 1969. "A New Subspecies of the Turtle *Graptemys nigrinoda* Cagle." *Copeia* 1969: 677–682.

Folt, B., and J. C. Godwin. 2013. "Status of Alligator Snapping Turtle (*Macrochelys temminckii*) in South Alabama with Comments on Its Distribution." *Chelonian Conservation and Biology* 12:211-217.

Fonnesbeck, C. J., and C. K. Dodd. 2003. "Estimation of Flattened Musk Turtle (*Sternotherus depressus*) Survival, Recapture, and Recovery Rate during and after a Disease Outbreak." *Journal of Herpetology* 37: 602–607.

Fritz, U., and P. Havas. 2007. "Checklist of Chelonians of the World." *Vertebrate Zoology* 57: 149–368.

Frost, D. R., and D. M. Hillis. 1990. "Species in Concept and Practice: Herpetological Applications." *Herpetologica* 46: 86–104.

Gaffney, E. S., G. E. Hooks, and V. P. Schneider. 2009. "New Material of North American Side-Necked Turtles (Pleurodira, Bothremydidae)." *American Museum Novitates* 3655: 1–26.

Gibbons, J. W. 1987. "Why Do Turtles Live So Long?" *Bioscience* 37: 262–269.

Guyer, C., and S. M. Hermann. 1997. "Patterns of Longevity of Gopher Tortoise (*Gopherus polyphemus*) Burrows: Implications for the Longleaf Pine-Wiregrass Ecosystem." *Chelonian Conservation and Biology* 2: 509–513.

Guyer, C., S. Glenos, S. M. Hermann, and J. Stober. 2011. "The Status of Gopher Tortoises (*Gopherus polyphemus*) in Alabama, with Special Reference to Three Important Conservation Properties." http://www.outdooralabama.com/sites/default/files/SWG%20Final%20Report%20-%20tortoise%20survey.pdf

Guyer, C., V. M. Johnson, and S. M. Hermann. 2012. "Social Behaviors of North American Tortoises." In *Biology and Conservation of North American Tortoises*, ed. D. C. Rostal, E. D. McCoy, and H. R. Mushinsky, 102-109. Johns Hopkins University Press.

Harper, R. M. 1943. *Forests of Alabama*. Rev. ed. Alabama Geological Survey Monograph 10. 230 pp.

Hauswaldt, J. S., and T. C. Glenn. 2005. "Population Genetics of the Diamondback Terrapin (*Malaclemys terrapin*)." *Molecular Ecology* 14: 723–732.

Hawkes, L. A., A. C. Broderick, M. S. Coyne, M. H. Godfrey, L.-F. Lopez-Jurado,

P. Lopez-Suarez, S. E. Merino, N. Varo-Cruz, and B. J. Godley. 2006. "Phenotypically Linked Dichotomy in Sea Turtle Foraging Requires Multiple Conservation Approaches." *Current Biology* 16: 990–995.

Hochscheid, S., B. J. Godley, A. C. Broderick, and R. P. Wilson. 1999. "Reptilian Diving: Highly Variable Dive Patterns in Green Turtle *Chelonia mydas.*" *Marine Ecology Progress Series* 105: 101–112.

Hochscheid, S., F. Bentivegna, and G. C. Hays. 2005. "First Records of Dive Durations for a Hibernating Sea Turtle." *Biology Letters* 1: 82–86.

Hughes, G. R., P. Luschi, R. Mencacci, and F. Papi. 1998. "The 7000-Km Journey of a Leatherback Tracked by Satellite." *Journal of Experimental Marine Biology and Ecology* 229: 209–217.

Iverson, J. B. 1977. "Reproduction in Freshwater and Terrestrial Turtles of North Florida." *Herpetologica* 33: 205–212.

Iverson, J. B., P. A. Meylan, and M. E. Seidel. 2008. "Testudines—Turtles." In *Scientific and Standard English Names of Amphibians and Reptiles of North America North of Mexico, with Comments Regarding Confidence in Our Understanding*, ed. B. I. Crother, 67–74. *SSAR Herpetological Circular* 37.

Jackson, C. G. Jr., and A. Ross. 1971. "The Occurrence of Barnacles on the Alligator Snapping Turtle, *Macroclemys temminckii* (Troost)." *Journal of Herpetology* 5: 188–189.

Jackson, D. R., and E. R. Milstrey. 1989. "The Fauna of Gopher Tortoise Burrows." In *Proceedings of the Gopher Tortoise Relocation Symposium*, ed. J. Diemer, J. Landers, D. Jackson, J. Layne, and D. Woods, 86–98. Florida Game and Freshwater Fish Commission, Nongame Wildlife Program, Technical Report Number 5, Tallahassee.

James, M. C., K. Martin, and P. H. Dutton. 2004. "Hybridization between a Green Turtle, *Chelonia mydas*, a Loggerhead Turtle, *Caretta caretta*, and the First Record of a Green Turtle in Atlantic Canada." *Canadian Field Naturalist* 118: 579–582.

Janzen, F. J., and G. L. Paukstis. 1991. "A Preliminary Test of the Adaptive Significance of Environmental Sex Determination in Reptiles." *Evolution* 45: 435–440.

Johnson, A. J., A. P. Pressier, J. F. X. Wellehan, A. Childress, T. M. Norton, N. L. Stedman, D. C. Bloom, W. Belzer, V. R. Titus, R. Wagner, J. W. Brooks, J. Spratt, and E. R. Jacobson. 2008b. "Ranavirus Infection of Free-Ranging and Captive Box Turtles and Tortoises in the United States." *Journal of Wildlife Disease* 44: 851–863.

Johnson, V. M., C. Guyer, and M. D. Boglioli. 2008a. "Phenology of Attempted Matings in Gopher Tortoises." *Copeia* 2008: 490–495.

Johnson, V. M., C. Guyer, S. M. Hermann, and W. K. Michener. 2009. "Patterns of Dispersion and Burrow Use Support Scramble Competition Polygyny in *Gopherus polyphemus.*" *Herpetologica* 65: 214–218.

Kaczor S. A., and D. C. Harnett. 1990. "Gopher Tortoise (*Gopherus polyphemus*) Effects on Soils and Vegetation in a Florida Sandhill Community." *American Midland Naturalist* 123: 100–111.

Krenz, J. G., G.J.P. Naylor, H. B. Shaffer, and F. J. Janzen. 2005. "Molecular Phylogenetics and Evolution of Turtles." *Molecular Phylogenetics and Evolution* 37: 178–191.

Lacefield, J. 2013. "Lost Worlds in Alabama Rocks: A Guide to the State's Ancient Life and Landscapes." The Alabama Museum of Natural History, Tuscaloosa, 276 pp.

Lahanas, P. 1982. "Aspects of the Life History of the Southern Black-Knobbed Sawback, *Graptemys nigrinoda delticola* Folkerts and Mount 1969. Master's thesis, Auburn University. 275 pp.

Lamb, T., and J. C. Avise. 1992. "Molecular and Population Genetic Aspects of Mitochondrial DNA Variability in the Diamondback Terrapin, *Malaclemys terrapin.*" *Journal of Heredity* 83: 262–269.

Lamb, T., and J. E. Lovich. 1990. "Morphometric Validation of the Striped Mud Turtles (*Kinosternon baurii*) in the Carolinas and Virginia." *Copeia* 1990: 613–618.

Langford, G. E. 1987. "Native American Legends." August House, Little Rock, AR.

Lara-Ruiz, P., G. G. Lopez, F. R. Santos, and L. S. Soares. 2006. "Extensive Hybridization in Hawksbill Turtles (*Eretmochelys imbricata*) Nesting in Brazil Revealed by MtDNA Analysis." *Conservation Genetics* 7: 773–781.

Le, M., C. J. Raxworthy, W. P. McCord, and L. Mertz. 2006. "A Molecular Phylogeny of Tortoises (Testudines: Testudinidae) Based on Mitochondrial and Nuclear Genes." *Molecular Phylogenetics and Evolution* 40: 517–531.

Leary, C. J., J. L. Dobie, T. M. Mann, P. S. Floyd, and D. H. Nelson. 2008. "*Pseudemys alabamensis* Baur 1893—Alabama Red-bellied Cooter, Alabama Red-bellied Turtle." Chelonian Research Monographs no. 5, Doi:10.3854/Crm.5.019 Alabamensis.V1.2008

Lindeman, P. V. 2008. "Evolution of Body Size in the Map Turtles and Sawbacks (Emydidae: Deirochelyinae: *Graptemys*)." *Herpetologica* 64: 32–46.

Liu, H., S. G. Platt, and C. K. Borg. 2004. "Seed Dispersal by the Florida Box Turtle (*Terrapene carolina bauri*) in Pine Rockland Forests of the Lower Florida Keys, United States." *Oecologia* 138: 539–546.

Lopez-Mendilaharsu, M., C.F.D. Rocha, A. Domingo, B. P. Wallace, and P. Miller. 2008. "Prolonged, Deep Dives by the Leatherback Turtle *Dermochelys coriacea*: Pushing Their Aerobic Dive Limits." JMBA2 Biodiversity Records, http//www.mba.ac.uk/jmba/jmba2biodiversityrecords.php.

Lovich, J. E., and T. Lamb. 1995. "Morphometric Similarity between the Turtles *Kinosternon subrubrum hippocrepis* and *K. baurii.*" *Journal of Herpetology* 29: 621–624.

Lovich, J. E., and C. J. McCoy. 1992. "Review of the *Graptemys pulchra* Group (Reptilia: Testudines: Emydidae), with Description of Two New Species." *Annals of the Carnegie Museum of Natural History* 61: 293–315.

Lutz, P. L., and J. A. Musick. 1997. "The Biology of Sea Turtles." *CRC Marine Biology Series.* V1. Boca Raton, FL: CRC Press.

Mahmoud, I. Y. 1969. "Comparative Ecology of the Kinosternid Turtles of Oklahoma." *Southwestern Naturalist* 14: 31–66.

Manire, C. A., L. Byrd, C. L. Therrien, and K. Martin. 2008. "Mating-Induced Ovulation in Loggerhead Sea Turtles, *Caretta caretta.*" *Zoo Biology* 27: 213–225.

McCoy, E. D., and H. R. Mushinsky. 2007. "Estimates of Minimum Patch Size Depend on the Method of Estimation and the Condition of the Habitat." *Ecology* 88: 1401–1407.

McCoy, E. D., H. R. Mushinsky, and J. Lindzey. 2006. "Declines of the Gopher Tortoise on Protected Lands." *Biological Conservation* 128: 120–127.

McGaugh, S. E., C. M. Eckerman, and F. J. Janzen. 2008. "Molecular Phylogeography of *Apalone spinifera.*" *Zoologica Scripta* 37: 289–304.

McKnight, T. L., and D. Hess. 2000. *Physical Geography: A Landscape Appreciation.* Upper Saddle River, NJ: Prentice-Hall.

Mills, H. H., and J. M. Kaye. 2001. "Drainage History of the Tennessee River: Review and New Metamorphic Quartz Gravel Locations." *Southeastern Geology* 40: 75–97.

Mirarchi, R. E., M. A. Bailey, T. M. Haggerty, and T. L. Best (eds.). 2004. *Alabama Wildlife. Vol. 3. Imperiled Amphibians, Reptiles, Birds, and Mammals.* Tuscaloosa: University of Alabama Press.

Moon, J. C., E. D. McCoy, H. R. Mushinsky, and S. A. Karl. 2006. "Multiple Paternity and Breeding System in the Gopher Tortoise, *Gopherus Polyphemus.*" *Journal of Heredity* 97: 150–157.

Mount, R. H. 1975. "Reptiles and Amphibians of Alabama." Alabama Agricultural Experiment Station, Auburn University, Auburn, AL.

Murray, C. M., C. D. McMahan, J. L. Dobie, and C. Guyer. 2014. "Cranial Variation amongst Independent Lineages of the Alligator Snapping Turtle (*Macrochelys Temminckii*)." *Journal of Zoological Systematics and Evolutionary Research* DOI: 10–1111/Jzs 12072.

Near, T. J., P. A. Meylan, and H. B. Shaffer. 2005. "Assessing Concordance of Fossil Calibration Points in Molecular Clock Studies: An Example Using Turtles." *American Naturalist* 165: 137–146.

Nelson, D. H., G. J. Langford, J. A. Borden, and W. M. Turner. 2009. "Reproductive and Hatchling Ecology of the Red-Bellied Cooter (*Pseudemys Alabamensis*): Implications for Conservation and Management." *Chelonian Conservation and Biology* 8: 66–73.

Nijs, J., and C. Navez. 1990. "Delayed Fertilization in the Striped Mud Turtle,

Kinosternon Baurii." *Bulletin of the Chicago Herpetological Society* 25: 146.

Osentoski, M. F., and T. Lamb. 1995. "Intraspecific Phylogeography of the Gopher Tortoise, *Gopherus polyphemus*: RFLP Analysis of Amplified Mtdna Segments." *Molecular Ecology* 4: 709–718.

Ott, J. A., M. K. Michener, and C. Guyer. 2003. "Patterns of Movement and Burrow Use in a Population of Gopher Tortoises (*Gopherus polyphemus*)." *Herpetologica* 59: 311–321.

Platt, S. G., H. Liu, and C. K. Borg. 2010. "Fire Ecology of the Florida Box Turtle (*Terrapene carolina bauri* Taylor) in Pine Rockland Forests of the Lower Florida Keys." *Natural Areas Journal* 30: 254–260.

Polovina, J. J., G. H. Balzs, E. A. Howell, D. M. Parker, M. P. Seki, and P. H. Dutton. 2004. "Forage and Migration Habitat of Loggerhead (*Caretta Caretta*) and Olive Ridley (*Lepidochelys olivacea*) Sea Turtles in Central North Pacific Ocean." *Fisheries Oceanography* 13: 36–51.

Pritchard, P.C.H. 1989. *The Alligator Snapping Turtle: Biology and Conservation*. Milwaukee Public Museum, Milwaukee, Wisconsin.

Renaud, M. L. 1995. "Movements and Submergence Patterns of Kemps Ridley Turtles (*Lepidochelys kempii*)." *Journal of Herpetology* 29: 370–374.

Romagosa, C. A., C. Guyer, and M. C. Wooten. 2009. "Contribution of the Live-Vertebrate Trade toward Taxonomic Homogenization." *Conservation Biology* 23: 1001–1007.

Roman, J, S. D. Santhuff, P. E. Moler, and B. W. Bowen. 1999. "Population structure and cryptic evolutionary units in the alligator snapping turtle." *Conservation Biology* 13: 135-142.

Ryan, T. J., C. A. Conner, B. A. Bouthitt, S. C. Sterrett, and C. M. Salsbury. 2008. "Movement and habitat use of two aquatic turtles (*Graptemys geographica* and *Trachemys scripta*) in an urban landscape." *Urban Ecosystems* 11: 213–225.

Schnuelle, K. A. 1997. "Demography, Diet, and Prey Availability of the Flattened Musk Turtle, *Sternotherus depressus*." Master's thesis, Auburn University. 82 pp.

Schofield, G., G. Katselidis, K. A. Pantis, J. D. Dimopoulos, and P. Hays. 2007. "Female-Female Aggressions: Structure of Interaction and Outcome in Loggerhead Sea Turtles." *Marine Ecology Progress Series* 336: 267–274.

Schwarzkopf, L., and R. J. Brooks. 1987. "Nest-Site Selection and Offspring Sex Ratio in Painted Turtles, *Chrysemys picta*." *Copeia* 1987: 53–61.

Seidel, M. E. 1994. "Morphometric Analysis and Taxonomy of Cooter and Red-bellied Turtles in the North American Genus *Pseudemys* (Emydidae)." *Chelonian Conservation and Biology* 1: 117–130.

Seidel, M. E. 2002. "Taxonomic Observations on Extant Species and Subspecies of Slider Turtles, Genus *Trachemys*." *Journal of Herpetology* 36: 285–292.

Shaffer, H. B., D. E. Starkey, and M. K. Fujita. 2008. "Molecular Insights into the Systematics of the Snapping Turtles (Chelydridae)." In *Biology of the Snapping Turtle (Chelydra serpentina)*, ed. R. J. Brooks, A. C. Steyermark, and M. S. Finkler. Baltimore: Johns Hopkins University Press.

Shealy, R. M. 1976. "The Natural History of the Alabama Map Turtle, *Graptemys pulchra* Baur, in Alabama." *Bulletin of the Florida State Museum of Biological Sciences* 21: 47–111.

Soltis, D. E., A. B. Morris, J. S. McLachlan, P. S. Manos, and P. S. Soltis. 2006. "Comparative Phylogeography of Unglaciated Eastern North America." *Molecular Ecology* 15: 4261–4293.

Spinks, P. Q., and H. B. Shaffer. 2009. "Conflicting Mitochondrial and Nuclear Phylogenies for the Widely Disjunct *Emys* (Testudines: Emydidae) Species Complex and What They Tell Us about Biogeography and Hybridization." *Systematic Biology* 58: 1–20.

Spinks, P. Q., R. C. Thomson, G. A. Lovely, and H. B. Shaffer. 2009. "Assessing What Is Needed to Resolve a Molecular Phylogeny: Simulations and Empirical Data from Emydid Turtles." *BMC Evolutionary Biology* 2009, 9: 56 Doi:10.1186/1471-2148-9-56.

Standora, E. A., and J. R. Spotila. 1985. "Temperature Dependent Sex Determination in Sea Turtles." *Copeia* 1985: 711–722.

Starkey, D. E., H. B. Shaffer, R. L. Burke, M.R.J. Forstner, J. B. Iverson, F. J. Janzen, A. G. J. Rhodin, and G. R. Ultsch. 2003. "Molecular Systematics, Phylogeography, and the Effects of Pleistocene Glaciations in the Painted Turtle (*Chrysemys picta*) Complex." *Evolution* 57: 119–128.

Steen, D. A., and J. P. Gibbs. 2004. "Effects of Roads on the Structure of Freshwater Turtle Populations." *Conservation Biology* 18: 1143–1148.

Steen, D. A., M. J. Aresco, S. G. Beilke, B. W. Compton, E. P. Condon, C. K. Dodd Jr., H. Forrester, J. W. Gibbons, J. L. Greene, G. Johnson, T. A. Langen, M. J. Oldham, D. N. Oxier, R. A. Saumure, F. W. Schueler, J. M. Sleeman, L. L. Smith, J. K. Tucker, and J. P. Gibbs. 2006. "Relative Vulnerability of Female Turtles to Road Mortality." *Animal Conservation* 9: 269–271.

Steen, D. A., S. C. Sterrett, S. A. Miller, and L. L. Smith. 2007. "Terrestrial Movements and Microhabitat Selection of Overwintering Subadult Eastern Mud Turtles (*Kinosternon subrubrum*) in Southwest Georgia." *Journal of Herpetology* 41: 532–535.

Steen, D. A., S. C. Sterrett, A. M. Heupel, and L. L. Smith. 2010. "Snapping Turtles, *Chelydra serpentina*, Overland Movements near the Southeastern Extent of Its Range." *Georgia Journal of Science* 68: 196–201.

Stephens, P. R., and J. J. Wiens. 2003. "Ecological Diversification and Phylogeny of Emydid Turtles." *Biological Journal of the Linnean Society* 79: 577–610.

Sterrett, S. C. 2009. "The Ecology and Influence of Land Use on River Turtles in Southwest Georgia." Master's thesis, University of Georgia. 113 pp.

Sterrett, S. C., L. L. Smith, S. W. Golladay, S. H. Schweitzer, and J. C. Maerz.

2010. "The Conservation Implications of Riparian Land Use on River Turtles." *Animal Conservation* 14: 38–46.

Sterrett, S. C., L. L. Smith, M. B. Howze, A. M. Heupel, and C. L. Atkinson. 2008. "*Pseudemys concinna* (Eastern River Cooter) Movement." *Herpetological Review* 39: 464.

Stone, P. A., J. L. Dobie, and R. P. Henry. 1992. "Cutaneous Surface Area and Bimodal Respiration in Soft-Shelled (*Trionyx spinifera*), Stinkpot (*Sternotherus odoratus*), and Mud Turtles (*Kinosternon subrubrum*)." *Physiological Zoology* 65: 311–330.

Styrsky, J. N., C. Guyer, H. Balbach, and A. Turkmen. 2010. "The Relationship between Burrow Abundance and Area as a Predictor of Gopher Tortoise Population Size." *Herpetologica* 66: 403–410.

Thomas, K. 1972. "The Annual Cycle of Reproduction of the Emydine Turtles, *Pseudemys floridana floridana* (Testudinata: Testudinidae) with Observations on Its Ecology." Master's thesis, Auburn University. 57 pp.

Thomas, T. M., M. C. Granatosky, J. R. Bourque, K. L. Krysko, P. E. Moler, T. Gamble, E. Suarez, E. Leone, K. M. Enge, and J. Roman. 2014. "Taxonomic Assessment of Alligator Snapping Turtles (Chelydridae: *Macrochelys*), with the Description of Two New Species from the Southeastern United State." *Zootaxa* 3786: 141–165.

Troëng, S., D. R. Evans, E. Harrison, and C. J. Lagueux. 2005. "Migration of Green Turtles *Chelonia mydas* from Tortuguero, Costa Rica." *Marine Biology* 148: 435–447.

Turtle Conservation Coalition (A. G. J. Rhodin, A. D. Walde, B. D. Horne, P. P. Van Dijk, T. Lanck, and R. Hudson, eds.). 2011. "Turtles in Trouble: the World's 25+ Most Endangered Tortoises and Freshwater Turtles—2011." Lunenburg, MA:IUCN/SSC Tortoise and Freshwater Turtle Specialist Group, Turtle Conservation Fund, Turtle Survival Alliance, Turtle Conservancy, Chelonian Research Foundation, Conservation International, Wildlife Conservation Society, and San Diego Zoo Global. 54 pp.

Ultsch, G. R., M. E. Carwile, C. E. Crocker, D. C. Jackson. 1999. "The Physiology of Hibernation among Painted Turtles: The Eastern Painted Turtles *Chrysemys picta picta*." *Physiological and Biochemical Zoology* 72: 493–501.

US Department of Interior. 2008. "Alabama Sea Turtle Conservation Manual." US Fish and Wildlife Service, Bon Secour National Wildlife Refuge, Gulf Shores, AL.

US Fish and Wildlife Service. 1987. "Determination of Threatened Status for the Flattened Musk Turtle *Sternotherus depressus*." *Federal Register* 52: 22418–22430.

Vogt, R. C. 1993. "Systematics of the False Map Turtles (*Graptemys pseudogeographica* Complex: Reptilia, Testudines, Emydidae)." *Annals of the Carnegie Museum* 62: 1–46.

Walker, D., G. Orti, and J. C. Avise. 1998. "Phylogenetic Distinctiveness of a Threatened Aquatic Turtle (*Sternotherus depressus*)." *Conservation Biology* 12: 639–645.

Walker, D., and J. C. Avise. 1998. "Principles of Phylogeography as Illustrated by Freshwater and Terrestrial Turtles in the Southeastern United States." *Annual Review of Ecology and Systematics* 29: 23–58.

Weisrock, D. W., and F. J. Janzen. 2000. "Comparative Molecular Phylogeography of North American Softshell Turtles (*Apalone*): Implications for Regional and Wide-Scale Historical Evolutionary Forces." *Molecular Phylogenetics and Evolution* 14: 152–164.

Wibbels, T. 2010. "The Diamond-backed Terrapin in Alabama: Causes for Decline and Strategy for Recovery." http://www.outdooralabama.com/sites/default/files/Diamondback%20Terrapin%20Final%20Report.pdf

Zbinden, J. A., C. R. Largiadér, F. Leippert, D. Margaritoulis, and R. Arlettaz. 2007. "High Frequency of Multiple Paternity in the Largest Rookery of Mediterranean Loggerhead Sea Turtles." *Molecular Ecology* 16: 3703–3711.

About the Authors

CRAIG GUYER was born on August 6, 1952, in Los Angeles to parents who migrated to several towns across Southern California before settling on a ranch on the outskirts of Oceanside. Although now decimated by high-density housing, during his childhood this area was remote, covered with coastal sage scrub vegetation, and home to healthy populations of amphibians and reptiles. So when he reached third grade, a stage when many children develop a fascination with wildlife, his backyard provided daily opportunities for exploration, an interest that he has never outgrown. Craig attended Humboldt State University in Northern California, graduating with a basic degree in 1975. He then moved to Idaho State University, completing a master's degree in 1978 that examined homing behavior in sagebrush lizards and short-horned lizards. He entered the lab of Dr. Jay. M. Savage, then at the University of Southern California but later chair of the Department of Biological Sciences at the University of Miami, for doctoral work. This career step introduced Craig to the diverse herpetofauna of Costa Rica, where he continues to maintain active research. He accepted a position at Auburn University in 1987, replacing the retired Robert H. Mount as curator of herpetology. Over the years the Guyer lab has maintained the premiere collection of Alabama amphibians and reptiles, provided research vital to the conservation of Alabama's amphibians and reptiles, and taught undergraduate students about global patterns of vertebrate biodiversity. He is the 2013 recipient of the Meritorious Teaching Award given by the Society for the Study of Amphibians and Reptiles and is the 2013–2016 Scharnagel Professor of Biological Sciences in the College of Science and Mathematics at Auburn University.

MARK A. BAILEY was born on July 20, 1961, in Birmingham, Alabama. His fascination with reptiles and amphibians began around age nine when he was given Hobart M. Smith's *Snakes as Pets*, and his formative years were spent roaming the woods of the upper Black Warrior River watershed, bringing home more "pets" than a conservation-minded herpetologist should probably confess. He was fortunate to have parents who tolerated and supported his interests. He attended public and private schools in Huffman, Palmerdale, and Pinson, Alabama. After graduating from high school in 1979 he entered Auburn University, where he completed a BS degree in biology in 1984 and a MS degree in zoology in 1988, with Robert H. Mount as major professor. From 1986 to 1988, he was assistant curator of the Auburn University Museum's herpetology collection. Immediately following graduation he worked as a biological technician at Conecuh National Forest, and in 1989 he was hired as a zoologist for The Nature Conservancy's new Alabama Natural Heritage Program. There he conducted statewide reptile and amphibian field surveys, conservation plans, and initiated the Alabama Herp Atlas Project. In 1998 he left the Heritage Program and with his wife, Karan, founded Conservation Southeast, a consulting firm they continue to operate from Andalusia. He has served on the advisory board of the Auburn University School of Forestry and Wildlife Science, and is a past president of the Alabama chapter of The Wildlife Society. He has served as the Alabama state representative to the Gopher Tortoise Council, and he continues to serve as a director of the Alabama Wildlife Federation. He received the Governor's Conservation Achievement Award for Wildlife Biologist of the Year in 2007. Mark has been involved in Partners for Amphibian and Reptile Conservation (PARC) and the Alabama chapter (ALAPARC) since their inception and was lead author on the 2006 PARC publication *Management Guidelines for Amphibians and Reptiles of the Southeastern United States.*

ROBERT H. MOUNT was born on December 25, 1931, in Lewisburg, Tennessee. Following the death of his mother four years later, he was cared for by his father and great aunt until his aunt died in 1937. He then went to live with his maternal aunt and uncle in Waynesboro, Tennessee. His father visited him weekly during that period and taught him to appreciate the natural world.

Waynesboro, with its crystal clear creeks and surrounding forests teeming with wildlife, was an idyllic place perfectly suited for a budding naturalist. Robert's father remarried when he was ten, and he, with his father and adoring stepmother, settled in Jackson, Tennessee, until moving to Albany, Georgia, where Robert attended high school. Following his graduation, he enrolled at Alabama Polytechnic Institute, now Auburn University, where he received BS and MS degrees. He then served in the Army in the United States and Korea as a medical entomologist for two years, after which he enrolled at the University of Florida as a PhD student in zoology under the direction of the late Dr. Archie Carr. Following his graduation he taught at Montevallo College in central Alabama, where he began his research on the herpetofauna of the state. Dr. Mount accepted a position at Auburn University in 1966, where he continued his research, ably assisted by colleagues Drs. Dan Speake, James Dobie, and George Folkerts, and his many graduate students, culminating in the 1975 publication of *The Reptiles and Amphibians of Alabama*. Mount held offices in a number of professional and environmental organizations and received several awards, including a Dudley Beaumont Fellowship, J. Kelly Mosley Award, Gopher Tortoise Award, Outstanding Environmental Writer Award by The Alabama Environmental Council, and a Lifetime Environmental Achievement Award by The Southern Appalachian Mountain Coalition. He retired in 1987 and in 2001 began writing newspaper columns, which he continues to do today.

Index